101 intriguing theories

Surendra Verma

This book is copyright material and must not be copied, reproduced, transferred, distributed, leased, licensed or publicly performed or used in any way except as specifically permitted in writing by the publishers, as allowed under the terms and conditions under which it was purchased or as strictly permitted by applicable copyright laws. Any unauthorized distribution or use of this text may be a direct infringement of the author's and publisher's rights and those responsible may be liable in law accordingly.

ISBN-13: 978-1975768416
ISBN-10: 1975768418

Copyright © Surendra Verma 2017

First published in paperback in 2017
Also available as an eboook

List of stories

Introduction   7

1. Abiogenic petroleum origin: Petroleum paradise   9
2. Alien abduction: Delusion or real?   11
3. Alien rain: It's raining aliens, not cats and dogs   14
4. Alien 'rods': Invisible creatures caught on camera   16
5. Ancient astronauts: Chariots of the gullible, not the gods   18
6. Anthropic principle: A scientific attempt to prove the divine   20
7. Apocalypse: Waiting for a celestial cataclysm   22
8. Archimedes' death ray: A burning tale from the past   24
9. Astrology: The planets of our lives   26
10. Ball lightning: A glowing, free-floating ball of light   29
11. Bermuda Triangle: The manufactured mystery of a deadly triangle   31
12. Bible code: Deciphering God's messages   33
13. Biological transmutations: Playing arithmetical games with atomic numbers   35
14. Biorhythms: Without rhyme or reason   37
15. Bode's law: A law of science or a remarkable coincidence?   39
16. Chiropractic: The real value of Palmer's 'discovery'   42
17. Climate scepticism: Myths can't change the reality   44
18. Cold fusion: Nuclear fusion in a glass jar   48
19. Common sense and science: A complex relationship   49
20. Consciousness: Explaining the unexplainable?   51
21. Cosmic collision: Planets colliding on the edge of science   54
22. Crop circles: Reaping by humans, aliens or nature?   56
23. Cryptozoology: Searching for snarks   59
24. Crystal healing: It's all in the mind   60
25. Delphic oracle: Inhaling vapours and prophesising   62
26. What really killed our lovable leapin' lizards   64

27. Dowsing: Does it work? *66*
28. Drake's equation: Sheer speculation *68*
29. Electromagnetic fields and health: Power line paranoia and mobile phone mania *71*
30. Energy from antimatter: Propelling fictional starships *74*
31. Evolutionary psychology: Can evolution explain how we think and behave? *76*
32. Extrasensory perception: Perception beyond belief *78*
33. Face on Mars: Unmasked! *81*
34. Flat Earth: No one has ever fallen off the edge *84*
35. Four elements: Aristotle's gift *86*
36. Geocentricity: Neither science nor religion *88*
37. Graphology: Handwriting as character *91*
38. Great Wall of China: The Moon myth *92*
39. Hollow Earth: Aliens in the deep *93*
40. Homeopathy: A medical breakthrough or a major blunder? *95*
41. Intelligent aliens: Why aren't they here? *97*
42. Intelligent design: Science or nonsense? *98*
43. Intercessory prayer: The power of distant prayer *101*
44. Killer asteroids: Should we lose sleep over asteroid threat? *103*
45. Kirlian photography: Images of fringes, not halos *105*
46. 'Left brain' and 'right brain' myth *107*
47. Litre: It's time you met Monsieur Litre and Mademoiselle Millie *110*
48. Loch Ness Monster: We see what we want to see *111*
49. Magnetic therapy: Attractive claims, but sham benefits *113*
50. Martian life: Smart Martians and silly scientists *114*
51. Mayan calendar: The end of a cycle *117*
52. Memes: Evolution by imitation *119*
53. Mesmerism: 'The art of increasing the imagination by degrees' *121*
54. Mirror matter: Looking glass worlds *123*
55. Moon hoax: Americans didn't land on the Moon, bat-men did *124*
56. Nanobots: The day of the self-replicating nanobots *126*
57. Near-death experiences: Near death, not near god *127*

58. Nocebo effect: The evil side of placebos   *129*
59. N-rays: Delusion, blunder or hoax?   *130*
60. Numerology: Superstition by numbers   *131*
61. Orion mystery: The Egyptian 'heaven on earth'   *133*
62. Out-of-body experiences: Out of body or out of mind?   *134*
63. Palmistry: Your future world in the palm of your hand   *136*
64. Panspermia: Life from outer space   *137*
65. Paradigm: A hopelessly overused and abused word   *139*
66. Parallel universes: They have found gazillion copies of you   *141*
67. Perpetual motion machines: Running forever without energy   *143*
68. Phrenology: Divining character from bumps on the head   *145*
69. Piltdown man: A synonym for phoney science   *147*
70. Placebo effect: Is belief one of the most powerful medicines?   *148*
71. Planet Nibiru: A fictional planet to end the world   *150*
72. Planet X: The saga of an unknown planet   *152*
73. Pole reversal and pole shift: Not the end of the world   *154*
74. Polywater: Contaminated with silica and silliness   *156*
75. Psychoanalysis: Still on the therapist's couch   *157*
76. Psychokinesis: Mind over matter   *158*
77. Pyramid power: A free way to sharpen your razor blades (if you use them)   *160*
78. Quantum healing: The non-physics of holistic healing   *162*
79. Quantum mysticism: Can quantum mechanics save your soul?   *164*
80. Rorschach inkblot test: Invisible and imprecise ink   *166*
81. Singularity: When humans will merge with machines   *168*
82. Sirius mystery: Fish gods of the star Sirius   *170*
83. Shroud of Turin: Unshrouding a mystery   *172*
84. Spontaneous human combustion: Can heavy drinking turn you into a 'crumbled black thing'?   *174*
85. Squaring the circle: It's impossible, indeed   *176*
86. Star of Bethlehem: The mystery of a supernatural star   *177*
87. Synchronicity: Coincidences – remarkable or random?   *180*
88. Tachyons: Faster than fact   *182*
89. Teleportation: Beam me up, Scotty!   *183*

90. Ten per cent brain myth: Untapped resources of the brain *185*
91. Tesla's death rays: Powerful enough to destroy 10,000 aeroplanes *187*
92. Time reversal: Can time go backwards? *190*
93. Time travel: Travelling back and forth in time *192*
94. Trepanation: You need it like a hole in the head *194*
95. Tunguska event: The riddle of a fireball *196*
96. UFOs: Do they exist? Try the Santa Claus Hypothesis *199*
97. Vitamin C: Cold comfort *202*
98. Wormholes: A tunnel from here to eternity *203*
99. 'Wow' signal: A missed phone call from ET? *205*
100. Zero-point energy: Not for harnessing, either by scientists or 'energy healers' *208*
101. Zoo hypothesis: Alien big brother is watching us all *209*

Appendix 1: Science and pseudoscience: Separating the wheat from the chaff *211*
Appendix 2: Beliefs in bizarre: Why are we suckers for weird beliefs? *214*

About the author *217*

Introduction

'CREATE A BELIEF in the theory and the facts will create themselves,' writes the American psychologist Joseph Jastrow in his book, *Wish and Wisdom: Episodes in the Vagaries of Belief* (1935). If you believe, say, in the 'flat-earth theory', search the Internet and you would have a vast collection of 'facts' to support the theory. You would even find 'historical facts' to show that your theory is credible because it has endured for centuries. You would also learn to argue that Apollo moon landings and the photos of the spherical Earth from space were faked: a big conspiracy to keep us all in the dark. Mark Twain's warning that a lie can travel halfway around the world before the truth gets its shoes on so aptly applies to the Internet.

This little book is not about the vagaries of belief and why we believe in weird things; it outlines 101 intriguing proposition, theories and other things from the borderlands of science. Not all concepts presented here could be dismissed as crackpot ideas; some have been proposed by respectable scientists and one day may even become mainstream scientific theories.

Appendix 1, 'Separating the wheat from the chaff', explains the difference between methods of credible scientists and crank pseudoscientists.

Appendix 2, 'Beliefs in bizarre', describes why are we suckers for weird beliefs.

Happy reading (with a touch of scepticism)!

Abiogenic petroleum origin

1

Petroleum paradise

THOMAS GOLD, THE brilliant but maverick Austrian-American scientist who died in 2004, is remembered for his unconventional theories. 'Gold's theories are always original, always brilliant, usually controversial – and usually right,' writes the renowned physicist Freeman Dyson in the foreword to Gold's last book, *The Deep Hot Biosphere: The Myth of Fossil Fuels* (2001). In this book, Gold states the case for his most controversial theory – he calls it his 'heretical views' – that contradicts the conventional wisdom that coal, crude oil and natural gas are fossil fuels, the fossilised remains of plants and animals that died millions of years of ago. On the contrary, says Gold, these resources are constantly being manufactured deep in the earth by natural processes from the initial materials that formed our planet. His abiogenic (not biological in origin) theory of petroleum origin ensures a virtually inexhaustible supply of petroleum and natural gas.

The conventional theory, on the other hand, holds that fossil fuels are residues of dead creatures who were buried in the sediments of inland seas or coastal marine basins. As this organic matter sank deeper into the earth under accumulated sediments it was subjected to increasing temperatures and pressures and underwent chemical reactions that distilled it into hydrocarbons. As the process takes millions of years, supplies of fossil fuels are not inexhaustible. Gold believes none of this. He says so-called fossil fuels were not produced from the decomposition of fossils, but by bacteria that are widespread at depth in the crust of Earth. This deep hot biosphere covers the entire crust down to a depth of several kilometres or miles. Unlike surface life, which is fed by photosynthesis (the process by which plants, algae and certain bacteria convert sunlight to chemical energy), the deep life is fed directly by chemical energy. This deep hot biosphere – a bacterial biosphere greater in mass than all life forms on the surface – has been flourishing on these resources for billions of years.

Extremophiles are bacteria that thrive in conditions that would kill other creatures – in deep-sea hydrothermal vents, rock chimneys that grow above volcanic vents on the sea floor, through which erupt hot, mineral-rich fluids; inside rocks buried kilometres below the Earth's surface where there is no oxygen and no organic food; in frozen Antarctic sea water; or in acidic, alkaline or saline environments. Recent discoveries of extremophiles far deeper in the Earth's crust than previously believed support Gold's theory.

Most geologists reject Gold's ideas on the grounds that the presence of biological markers in petroleum supports the conventional theory. The discovery of extremophiles supports Gold's theory by showing that biological matter in petroleum result from a bacterial action on abiogenic methane. Does Gold's theory promise a fuel paradise or a fool's paradise?

In 1960 Gold suggested another interesting idea: that life on Earth began when it was infected by micro-organisms in the garbage left over by extraterrestrial visitors. He imagined that interstellar visitors forgot to clean up after having a picnic on our planet. Do piles of garbage left in picnic spots around the world hint we have inherited this bad habit from our interstellar ancestors?

Alien abduction

2

Delusion or real?

ALIEN ABDUCTION STORIES abound in the media and on the Internet. Most stories are strikingly similar and follow a consistent pattern. If you're ever abducted by aliens:

Encounter: You will be either in your bed or car, usually at night. If you are in your bed, you suddenly wake up and see non-human figures coming through the walls of your room or standing near your bed. You may even see a spaceship outside your window. If you are in your car, you see the car being pulled to the side by a bright object. You see strange beings exiting from a spaceship and coming towards you.

You are unable to move or speak. You see flashing lights and hear buzzing sounds.

The aliens are usually about 120 centimetres (4 feet) tall. Their bodies are thin and spindly with huge heads, slanted wrap-around black eyes, grey skin and no hair or nose. They don't talk but communicate with you telepathically.

Abduction: You are taken to the spaceship. You don't walk; you either float or are carried up by a beam of light. The inside of the spaceship looks like a high-tech doctor's examination room full of tables on which other humans lie.

Examination: You are subjected to a painful medical examination. The aliens probe you by inserting instruments into virtually every part of the body. Sometimes they place tiny implants in your body, especially the nose. No person has ever been found with an actual implant. Most abductees claim that they lost the nose implant when they sneezed.

Tour: The aliens may give you a tour of their spaceship or show some alien artefacts. No person has ever brought back any extraterrestrial souvenirs from their flying saucer tour.

Return: It's all over within a short time. You are back in your bed or car. You are confused and scared. You may find puzzling 'surgical' scars or cuts on your body and you don't remember how you got them.

The alien-abduction phenomenon began with the case of Betty and Barney Hill. Late one night in 1961, when Betty and her husband Barney were returning to their New Hampshire home from a holiday in Canada, they noticed a bright light following their car. The light became brighter and brighter until it was clearly visible as a spaceship. Their car stalled and the spaceship landed nearby. The aliens came out and took them inside the spaceship, where they were subjected to a medical examination. Betty was given a brief tour of the spaceship and when she asked where the aliens were from she was shown a star map.

When they 'awoke', the couple continued their drive home, but without any memories of the incident. Weeks afterwards, they started experiencing nightmares and went to see a psychiatrist. Under hypnosis they 'remembered' having been abducted by aliens and subjected to painful probing of their bodies; Betty also drew the star map shown by the aliens. A few years later, a UFO researcher claimed that Betty's vague star map resembled Zeta Reticuli, a binary star system located in the constellation Reticulum, 39 light years away.

Betty and Barney's story became highly popular when John Fuller told it in his book, *The Interrupted Journey* (1966). It was later made into the television movie *The UFO Incident* (1975). The book and the movie started a new genre, and the publishing and entertainment industries continue to milk this cash cow.

People who believe in alien abduction have a tendency to fantasise and to hold to unusual beliefs and ideas. They also believe in things like ESP, astrology, tarot, channelling, auras and crystal therapy. Susan Clancy, author of *Abducted: How People Come to Believe They Were Kidnapped by Aliens* (2005), says these people are not crazy, but they have in common a rash of disturbing experiences for which they are seeking an explanation. For them, alien abduction is the best fit. 'Many of us long for contact with the divine, and aliens are a way of coming to terms with the conflict between science and religion,' she says. 'I agree with Jung: extraterrestrials are technological angels.'

Some psychologists have tied the phenomenon to sleep paralysis, a condition where the usual separation between sleep and wakefulness gets out of synchronisation. Sleep occurs when the body is in the dream phase of sleep and it disconnects from the brain. The brain is either awake or semi-awake but the body cannot move. At that point, the sleeper often 'sees' shadowy creatures, 'experiences' levitation and 'feels' painful sensations throughout the body, explains Kazuhiko Fukuda, a Japanese psychologist. Such experiences match with the accounts of people claiming to be victims of alien abduction.

There are other psychological explanations as well. Some psychologists believe that alien abductions and other mystical and psychic experiences may be linked to excessive bursts of electrical activity in the temporal lobes. These lobes – one on each side of the brain, located near the ears – control hearing, speech and memory. Some argue that alien abduction may be disguised memories of sexual abuse. Or, they may be false memories. People can and do make powerful memories and these memories can take on a life of their own. All abduction stories have been recalled under hypnosis. Hypnosis makes people susceptible to creating memories of things that were suggested to them or things they just imagined.

US psychologist David V. Forrest offers two medical hypotheses. His 'strong' hypothesis is that the abductees are recovering memories of actual surgery – the memories may be an actual recall of the operating room before losing consciousness or they could be memories from childhood filtered through childhood amnesia. His 'weak' hypothesis is that abductees are confabulating media conceptions of aliens with images of surgical and medical procedures generally, images that they may or may not have experienced personally.

Whatever may be the right explanation for accounts of alien abduction, why are their stories so similar? Anthropologists point out that individual illusionary experiences conform to cultural patterns. Alien abductions occur mainly in the United States where people are familiar with alien references through supermarket tabloids, books, movies and TV shows. Abduction accounts became appearing from 1962 when alien abduction also began appearing on TV and the movies. During the witchcraft craze in medieval Europe, for example, many people reported being carried away by witches on broomsticks and being seduced by demons. Today's counterparts talk of being picked up by flying saucers and being forced to perform various forms of unwilling sex by their bug-eyed kidnappers.

3
It's raining aliens, not cats and dogs

IF INTELLIGENT ALIENS are visiting Earth by UFOs, microbial aliens are travelling by rain to visit Earth, at least a tiny part of it in Kerala in south-western India. On the morning of 25 July 2001, residents of a small town in the Kottayam district heard a loud sonic boom accompanied by lightning. What followed was a three-hour spell of heavy monsoon rain. For about fifteen minutes the rain turned blood red, which eyewitnesses claim was falling as scarlet sheets. Many people in the street found their clothes turned pink. Some even noticed that the rain burned leaves on trees. Similar blood-red rain bursts during normal rains continued sporadically there and in other areas along the coast for about two months.

Explanations provided by local experts included: a burning meteorite threw out red dust which came down as red rain; fine dust blown from Arabian deserts got mixed with monsoon rains; red rain particles were possibly fungal spores from trees; and a fine mist of blood cells produced by a meteor striking a high-flying flock of bats.

The red rain would have been forgotten completely if Godfrey Louis, a solid-state physicist at Mahatma Gandhi University in Kottayam, had not decided to investigate this mysterious phenomenon with his student Santhosh Kumar. From various news reports and other sources, the researchers compiled a list of 124 incidences of red rain, and collected samples of red rainwater from various places more than 100 kilometres (62 miles) apart.

Their analysis of the rainwater, published in 2006, shows that the particles are microscopic in size, almost transparent red in colour and well dispersed in the water. The particles are clearly not sand but have an appearance similar to single-celled organisms. One chemical analysis has shown that they contain about 50 per cent carbon, 45 per cent oxygen and traces of sodium and iron. All this data is consistent with biological materials. But the cells lack any nucleus and DNA. Life as we know it must contain DNA; although some scientists say that this wouldn't necessarily be true of alien microbes. Are they alien microbes?

Louis and Kumar conclude that the sonic boom heard by many people before the rain was triggered by a meteorite. When it exploded in the upper atmosphere, it shed the embedded alien microbes, which were mixed with the clouds and then fell as rain. Another analysis in 2010 has shown an unusual pattern in the way cells changed colour under UV light – known as fluorescence behaviour – which they say suggests an extraterrestrial origin.

In the 1970s the maverick British scientist Fred Hoyle advanced the controversial idea that chemical building blocks of life are present in interstellar clouds. When these clouds collapse to form comets, they provide likely sites for the origin of life. Microbes multiply inside a comet, which has a warm, liquid interior. An impact of a comet about 3.8 billion years ago could have led to the start of life on Earth (*story 64*).

Was the red rain of Kerala terrestrial or alien? Even if terrestrial, how did it happen? Scientists are still debating, but the 'alien rain' has already found a home on thousands of alternative science websites and blogs. *X-Files* fans don't like to wait for scientific verdicts.

Alien 'rods'

4
Invisible creatures caught on camera

THE 'RODS' WERE first 'discovered' in 1994 when José Escamilla, an American UFO buff, was videotaping UFOs that had appeared in broad daylight near Roswell, the famous 'UFO site' in New Mexico (*story 96*). When reviewing the tape he saw streaks on the screen. He first thought that they were caused by flying insects or birds flying close to the video camera. When he reviewed the tape frame by frame, he realised that the streaks were not insects or birds but something else. His wife called them 'rods' because they resembled microbes she had seen under a microscope.

Escamilla rejected the phenomenon as an optical anomaly and concluded that the 'rods' were living organisms, probably of extraterrestrial origin. He continued to study these mysterious beings and appeared on many local radio and TV shows to publicise his 'discovery'. UFO and cryptozoology (*story 35*) enthusiasts latched onto the idea and soon the 'rods' found a place in their annals of extraterrestrial and unexplained phenomena.

'Rods' almost invariably appear only in photographs, films and videotapes, although Escamilla claims that they can be seen with the naked eye and speculates them to be anywhere from a few inches to a hundred or more feet in length. He also claims to have recorded many types of 'rods': the 'centipede' types, which have appendages across the torso; 'white rods', which have a ribbon-like appearance; and super-thin 'spears', which move very fast.

Bob Duhamel, an American amateur astronomer who has extensively studied the 'rods', has come to the conclusion that it was a 'shameless hoax'. He explains the 'rods' in terms of artefacts introduced to photographs by the nature of lenses, film (or charge-coupled devices, in the case of a digital camera) and shutters. One of the most common artefacts is lens flares, bright spots superimposed on the photograph by reflections of a bright object from the several surfaces of a compound lens. 'Inexperienced or perhaps gullible people often think they are UFOs or even angels,' he says. Another artefact is the fact that an object with high angular velocity will appear as a streak. 'When a blurred streak appears on a photograph most of us will see it as a fast-moving object; José Escamilla sees them as unidentified life forms,' he says.

Duhamel's explanation has failed to convince rods' buffs. Escamilla still promotes 'rods' on his website (www.roswellrods.com), and Duhamel has posted a challenge on his website (www.amsky.com) to 'prove that "rods" are not artefacts of photographic process (this proof must be capable of withstanding peer review in the legitimate scientific community).'

Benjamin Radford of *Skeptical Inquirer* magazine has the last word: 'In a nutshell, rods are to cryptozoology (or UFOs) what orbs are to ghosts.'

5
Chariots of the gullible, not the gods

SCIENTISTS ARE STILL arguing about the existence of microbial life on other worlds, not to speak of intelligent life; but in 1968 Erich von Daniken, a mild-mannered Swiss hotelier with no scientific training, published a book in which he claimed that our planet was visited by intelligent aliens in prehistoric times. Published in English in 1969 under the tile *Chariots of the Gods?*, the book became a bestseller and inspired a whole 'ancient astronauts' industry of countless books, documentaries and films.

The fad has almost run its course, but not completely. The book is still in print and tens of thousands of websites are now devoted to its startling but poorly researched ideas.

With the zeal of a UFO convert, von Daniken professed that we are the descendants of astronauts from outer space who came to Earth about 10,000 years ago; our gods are simply the astronauts that visited us in the past. He said that mythical stories of many lands, which are filled with godlike visitors from the sky arriving in fiery chariots, support this assertion.

He argued that 'unsolved mysteries of the past' such as the moving of large stones for the building of pyramids and the great carved heads of Easter Island could be explained by knowledge learned by our ancestors from visiting aliens. He suggested that ancient patterns of triangles, rectangles and trapezoids, huge spiders, monkeys, birds, fish and reptiles laid out with equal precision across 50 kilometres (31 miles) of Peru's Nazca plain were landing strips for 'ancient astronauts'. It's true some Nazca patterns are so large that they could only be recognised from the air, but there is a down-to-earth explanation for them: they were probably drawn by Nazca's astronomer-priests to mark the passage of seasons.

Archaeologists haves questioned von Daniken's research techniques and reasoning. They say that, of course, there are things which are still mysteries from the past, but von Daniken's freewheeling speculations based on bits and pieces of archaeological 'facts' have failed to present a cohesive and convincing argument.

6
A scientific attempt to prove the divine

THERE'S NOTHING SPECIAL about the conditions on this planet. There's no reason why things should be different anywhere else in the universe. In other words, one's location is unlikely to be special, or to put it bluntly, wherever or whenever we are, it's nothing special. This is known as the 'Copernicus principle' after Copernicus, who in 1543 demoted Earth to an ordinary unprivileged place in the cosmos.

In contrast, the anthropic principle (from Greek *anthropos*, 'human beings') maintains that human beings hold a special place in the universe. The fundamental laws of physics that govern the universe are not the result of chance but somehow fine-tuned to allow the existence of intelligent life. If, for example, the force of gravity was slightly different from it is now, there would be no Sun-like stars anywhere.

Simply put, we live in a Goldilocks universe ('Ahhh, this porridge is just right') because the universe has 'just right' conditions for the existence of life. The universe has to be the way it is.

The term 'anthropic principle' was proposed by Brandon Carter, a British cosmologist, in 1973 at a symposium in Poland commemorating Copernicus's 500th birthday. He suggested two versions of the principle: (1) the weak anthropic principle: the conditions in the universe are compatible with our existence; and (2) the strong anthropic principle: the universe must have those properties which would allow intelligent life within it at some stage. Other scientists have suggested different possible implications of the strong principle, including that the universe was 'designed' with the goal of sustaining human beings. This has been interpreted as evidence for a creator.

However, many cosmologists dismiss the anthropic principle as being non-science, because it makes no testable predictions. British cosmologist Stephen

Hawking (the author of *A Brief History of Time*) and Leonard Mlodinow write in *The Grand Design* (2010): 'The idea that the universe was designed to accommodate mankind appears in theologies and mythologies dating from thousands of years ago ... That is not the answer of modern science. As recent advances in cosmology suggest, the laws of gravity and quantum theory allow universes to appear simultaneously from nothing. Spontaneous creation is the reason there is something rather than nothing, why the universe exists, why we exist. It is not necessary to invoke God to light the blue touch paper and set the universe going.'

Hawking and Mlodinow also say that our universe seems to be one of many, each with different laws. 'The idea of multiverse is a consequence predicted by many theories in modern cosmology. If it is true, it reduces the strong anthropic principle to the weak one, putting the fine tunings of physical law on the same footing as the environmental factors.' This leads to only one conclusion: the universe is just one of many. It's not custom made for us.

Apocalypse

7

Waiting for a celestial cataclysm

THE PLANET EARTH has survived for 4,600 million years; no celestial cataclysm has succeeded in disintegrating it into cosmic dust. Similarly, life on this puny planet has dodged total extinction many times for 3,500 million years when battered by killer comets and asteroids from the sky and deadly megaquakes and supervolcanoes from the deep. For nearly two million years, humans have survived ice ages and other natural catastrophes. It's true that civilisations exist by geological consent and when nature suddenly withdraws this consent, rumbling earthquakes or oozing volcanoes wipe out vast swathes of human societies, but not humankind. Yet, our fascination with prophecies about the end of the world (or at least, the world without us) is endless.

Waiting for apocalypse seems like a weird preoccupation. It's not. The word apocalypse comes from a Greek word meaning uncover or reveal. In the Bible it refers to the Revelation to John, the last book of New Testament. The Apocalypse is the final battle between good and evil. This battle, prophesied to be fought at a place called Armageddon, will bring Earth's history to an end. Many theologians believe that the purpose of the Apocalypse was not to scare and depress people but to encourage them and carry them through difficult times.

The Qu'ran also talks about the Day of Judgment: On the Last Day when the sun rises from the west and Earth is rocked in her last convulsions humanity will come in broken bands to be shown their labours; whoever does an atom's weight of good shall see it, and whoever does an atom's weight of evil shall see it as well. The Last Day could come anytime.

The concept of the end of time also appears in Judaism, but not in Hinduism and Buddhism. In these religions time is perceived as cyclical, not linear. They believe in the cosmological doctrine of the repeated creation and destruction of the universe. Hindu scriptures, however, describe four cycles of eras, yugas. At the end of the fourth and the current cycle, kali yuga ('the age of unrighteousness'), the world shall be destroyed by a cosmic fire. When? It depends on who is interpreting (or misinterpreting) the texts: we could have been the victims of the demon kali in 2012 (on 21 December, the day cataclysmic events were supposed to occur according to the Mayan calendar (*story 51*) or we might have another 427,000 years to live.

The word apocalypse now means destruction on a catastrophic scale, and armageddon is synonymous with the end of the world. Whatever we do, it's inevitable that the history of the world will come to a close some day. That day could be billions of years away or it could be tomorrow. Now seems to be the perfect time when faith, science, current events and human frailty to seek affirmation in what we believe and fear are working together to create an ideal climate of seductive fascination with apocalypse.

You can be sceptical and dismiss all end-of-the-world ideas as pseudoscientific nonsense. If you're gullible and have no sceptical sense, there is danger that you would never distinguish good ideas from bad ones and constantly worry about the apocalypse.

Archimedes' death ray

8
A burning tale from the past

WHEN IN 214 BC the Roman general Claudius Marcellus besieged Syracuse, a Greek colony in Sicily, King Hieron of Syracuse persuaded Archimedes, his friend and relative, to prepare all kinds of ingenious offensive and defensive engines to be used in every kind of siege warfare. The Romans were panicky to see huge catapults, enormous grappling irons and cranes and big mirrors to set fire to the wooden ships. On seeing the destruction caused by a single man Marcellus said, 'Let us stop fighting against this geometrical Briareus, who outdoes the hundred-handed monsters of mythology'. However, the Roman army captured the city in 212 BC.

Two hundred years later the Greek historian Diodorus Siculus (literally 'Diodorus the Sicilian') wrote: 'When Marcellus now moved the vessels off as far as an arrow can fly, the old man [Archimedes; he was then 74 years old] then devised a hexagonal mirror, and at an appropriate distance from it set small quadrangular mirrors of the same type, which could be adjusted by metal plates and small hinges. This contrivance he set to catch the full rays of the sun at noon, both summer and winter, and eventually, by the reflection of the sun's rays in this, a fearsome fiery heat was kindled in the barges, and from the distance of an arrow's flight he reduced them to ashes. Thus did the old man, by his contrivances, vanquish Marcellus.' (Loeb Classical Library, Harvard University Press, 1980)

Siculus' account is probably the source of the enduring legend that Archimedes used mirrors to focus the Sun's rays to set fire to Roman ships besieging his home town.

The first attempt to prove the legend was made in 1747 by the famous French naturalist Georges Louis LeClerc de Buffon. He carried out elaborate experiments with 168 mirrors, each about the size of an A4 paper, and was able to burn a piece of wood about 50 metres (160 feet) away.

Centuries passed before another serious attempt was made to prove or disprove Archimedes. In 1973 Ioannis Sakkas, a Greek engineer, concluded that the 'burning mirror' of legend was not a large convex mirror because such a technology was beyond the technology of Archimedes' days. He thought the mirror was probably highly polished metal – most likely the shields of Roman soldiers. He ordered dozens of flat mirrors covered with a thin reflecting sheet of polished copper. Each was about the size of a Roman soldier's large rectangular shied, easily handled by one person. He then lined up 70 people holding these mirrors on a pier and directed them to focus on a boat about 18 metres (60 feet) away. The boat started smoking within a few seconds and was engulfed in flames.

Two experiments conducted by Discovery Channel's *Mythbusters* show in 2004 and 2006 failed to ignite a mock boat. Myth busted, according to the show. They suggest that flaming arrows and fireballs are much easier to create and Archimedes probably used these devices, not mirrors.

Dismiss the story of Archimedes' death ray as a myth? After all, not all ancient stories can be taken at face value.

Astrology

9

The planets of our lives

THE SKY WAS once full of portents. A comet or conjunction of planets filled ordinary people with terror. The perception of the influence of planets on human lives could make mighty emperors and generals look up to the heavens for guidance. Since Isaac Newton's gravity forced order on the wandering of planets and other heavenly bodies, astrology has been intellectually weightless. Yet, it has not lost its seductive powers to charm us.

Though astrology and astronomy had a starry relationship in the past, astrology is not a science. Obviously, astrologers disagree. They even assign astronomy a lower status as an offspring of astrology. Astronomy has made remarkable progress in the past few centuries, but astrology has hardly changed. Tell this to astrologers and they would cast the horoscope of their 'science': astrology is truth and truth never changes. But we know now that some ancient 'truths' are not eternal. For example, when astrologers say that the Sun is in Aries between 21 March and 21 April, they mean that the Sun, as seen from Earth, is in the same part of the sky as the constellation of Aries. They ignore the astronomical phenomenon of precession, the slow, conical motion of the Earth's axis of rotation which makes Earth wobble around its axis in a 25,800-year cycle. Precession makes Earth's position relative to the constellations of the zodiac change over centuries. The Sun is no longer in the constellation of Aries; it's in the constellation of Pisces. Shouldn't it affect you, if you are an Aries, or any other astrological sign?

The ancients believed that Earth was the centre of the universe, fixed and unmoving. They saw the planets generally moving relative to the fixed stars from west to east. But every once in a while a planet will slow down, stop and then reverse direction (retrograde). They explained this puzzling motion by saying that the gods were annoyed and therefore 'malefic'. How could the 'malefic' or other positions of planets influence our lives?

Nearly five centuries have passed since Nicolaus Copernicus displaced Earth from its throne and set it in motion. Astronomers have now discovered hundreds of planets beyond the solar system. In comparison, the planets of astrology are indeed puny and their influence on Earth insignificant.

Scientists talk about four forces that influence everything in the universe. Gravity is the long-range force: it holds chairs to the floor and planets in their orbits. The electromagnetic force is the attraction and repulsion between charged particles: it enables light bulbs to glow and lifts to rise. The strong force keeps atomic nuclei together: it binds together the protons and neutrons in an atomic nucleus. The weak force is also a kind of nuclear force: it causes elementary particles to shoot out of the atomic nucleus during the nuclear decay of such radioactive elements as uranium. The strengths of the forces vary widely. The strong force is one sextillion (1 with 36 zeros after it) times stronger than gravity, making gravity the weakest force. The strong force may be strong but its reach is limited to the atomic nucleus.

Gravity influences us, but the tug of distant planets on you is smaller than the gravitational pull of this book. The American astronomer Carl Sagan once pointed out that the gravitational pull of the midwife would have far greater influence at a child's birth than the pull of a planet.

Does astrology work? Many major studies have tried to answer this question, but none has found any proof that the astrological sign you were born under influences the person you become. The largest test of astrology ever taken – a statistical analysis of the birthdays of more than twenty million married people in England and Wales – has shown that their astrological sign has no impact on the probability of marrying and staying married to someone of any other sign. There is no such thing as zodiac signs and love compatibility. Googling 'love sign astrology' would produce hundreds of thousands of hits, but none of them is going to make you wiser in choosing your life partner.

Many studies, however, have shown that being born at certain times of year relates to a small but significantly increased risk of problems such as depression, schizophrenia and anorexia nervosa. People born in late winter and early spring in the northern hemisphere have a slightly higher risk of developing schizophrenia than those born at other times of the year. Similarly, northern hemisphere autumn birthdays are associated with an increase in the likelihood of suffering panic attacks. These are medical studies trying to find out how seasons affect mental health; in no way do they support astrology.

As a science astrology is silly; as an entertainment it's harmless; so is numerology (*story 60*).

Ball lightning

10

A glowing, free-floating ball of light

BALL LIGHTNING IS one of the nature's rarer phenomena and few photographs of it exist. It is also the least understood. Ball lightning has attracted the attention of scientists for two centuries, but it remains an enigma – dismissed by many as myth or an optical illusion. Here are three eyewitness accounts of ball lightning sightings.

One account describes a lightning ball as it entered through an open window in the pantry of a house in Johannesburg. 'It entered the kitchen around the corner, then sped out around another corner and into the hall where it hit the tin bucket with a clang. Certainly when we ran to check, the bucket was too hot to pick up and its paint had blistered.'

In another account, a white-grey lightning ball about 80 centimetres (30 inches) in diameter and with the glow of an incandescent lamp of 200 watts bounced on the head of a Russian teacher who was with her friends: 'It appeared as if from nowhere. We got frightened, squatted and connected our heads, creating a circle. The ball suddenly began to move over us in a circle and also moved up and down. Then it 'chose' my head and began to jump on it, up and down, like a ball. It made more than twenty jumps. It was soft as a bubble.'

One of the rare accounts of ball lightning witnessed by a respected scientist comes from the British radio astronomer R.C. Jennison, who encountered a lightning ball on an aircraft in 1963. He described it as a blue-white glowing sphere a little more than 20 centimetres (8 inches) in diameter which emerged from the pilot's cabin and passed down the aisle of the aircraft at a short distance from him, maintaining the same height and course for the whole distance over which it could be observed. It then passed through the metal wall of the aircraft.

From such accounts, scientists have painted a picture of this bizarre phenomenon, which is always observed during stormy weather. A lightning ball is usually seen as a free-floating, luminous sphere that shines for a few seconds to a few minutes before it either explodes with a sharp bang or flicks out in silence. It can be almost any colour, sometimes even a combination, but green and violet are rare. Its size varies from a small ball to a giant globe several metres in diameter. It may suddenly appear in the air, or even from holes in the ground, sewers and ditches. It usually moves horizontally in the air about a metre (40 inches) above the ground, but can climb utility poles and then dart along power or telephone lines. It can even dive chimneys and squeeze through spaces much smaller than itself, but it never changes its size. It seems cool to the touch, but it may destroy electrical equipment, melt glass, ignite fires and scorch woods or singe people and animals. Sometimes a hissing or crackling noise can be heard. It may leave behind a sharp and repugnant smell, resembling ozone.

Ball lightning has been the subject of serious scientific research since the early 19th century, but no consensus theory has yet emerged to explain what causes it, or even exactly what it is. One of the popular theories says that a lightning ball is a sphere of plasma, or a hot gas of electrons and positively charged ions. Another theory says that when lightning strikes soil, it turns silica in the soil to pure silicon vapour. As the hot vapour cools, the silicon condenses into a floating ball of silicon aerosol held together by electrical charges. The chemical energy stored in silicon is slowly released as heat and light. Because the ball would become visible only over the latter part of its lifetime, it would appear to materialise out of thin air after a lightning strike.

So simple, yet so amazing! Unfortunately, most of us will never see it (about 1 per cent of the population will see ball lightning in a lifetime). Some scientists even reject that ball lightning exists. They say that there is no reliable photograph of ball lightning which is not explainable otherwise; all known forms of electric discharge can be reproduced in the lab, but not ball lightning.

11

The manufactured mystery of a deadly triangle

THE BERMUDA TRIANGLE is a heavily travelled stretch of sea between the Bahamas, Bermuda and the east coast of the United States. In 1974 Charles Berlitz, grandson of the language school-founder, made a sensational claim in his book, *The Bermuda Triangle*, that numerous ships and aeroplanes had disappeared under mysterious and unexplainable circumstances while traversing the triangle. The book – which *Time* magazine described as 'a hodgepodge of half-truths, unsubstantiated reports and unsubstantial science' – immediately made its way onto the bestseller lists.

Berlitz cited many imaginative explanations for the sudden disappearances of ships and aeroplanes: kidnapped by UFOs or sucked in by 'magnetic vortices' created by powerful energy sources produced long ago by an ancient civilisation on the lost continent of Atlantis that sank to the bottom of the Atlantic Ocean.

Three years later Berlitz came up with a far more speculative explanation in his new book, *Without a Trace*: he had found a great pyramid in the deadly triangle which somehow was responsible for all the disappearances.

The truth of the story is simply humdrum. There are no geological anomalies in that part of the ocean, and in 1975 shipping records of Lloyd's of London showed that 428 vessels had been reported missing throughout the world since 1955, but there was no evidence to support the claims that the Bermuda Triangle had more losses than elsewhere.

The handful of disappearances of ships in the triangle could be simply explained by stormy weather occurring over the warm water in that area. The triangle is also one of two areas where the magnetic compass points to true north, which may confuse some navigators.

The once-famous Bermuda Triangle mystery was nothing more than a fad, yet many people still believe in it for the simple 'logical' reason that they have heard so much about it and therefore there must be some truth in it. Some have even described it as a manufactured mystery – a hoax – and hoaxes are built on our gullibility.

Bible code

12
Deciphering God's messages

FOR 3,000 YEARS a code in the Torah, the first five books of the Jewish Bible which correspond in some ways to the Christian Old Testament, has remained hidden. Now it has been unlocked by a computer. The code predicts events – such as the Second World War, the Holocaust, the atomic blast at Hiroshima and the landing on the moon – that happened thousands of years after the Bible was written. This extraordinary claim has been made in *The Bible Code*, a book by an American journalist Michael Drosnin, published in 1997. It has been a phenomenal success and has started a 'Bible codes' industry of books, TV documentaries, websites and at least one movie.

The code was 'deciphered', the book claims, by Eliyahu Rips, an Israeli mathematician. The entire Torah has some 304,805 Hebrew letters. The code is revealed by searching for equidistant letter sequences. For example, start with any letter and then take every *n*th letter (*n* can be any number you wish) in the book. Print out that string of letters in a block of type (which will resemble a larger version of popular find-the-words puzzles). The grid could then be searched from left to right, right to left, up and down and diagonally in any direction for any interesting pattern of words. As one of the examples, Drosnin's book shows that the name of Yitzhak Rabin, the Israeli prime minister who was assassinated in 1995, was in close proximity to the words 'assassin that will assassinate'. After the publication of the book, there were many claims that these 'messages' in the Jewish Bible were not coincidences but deliberately put there by God.

Mathematicians have described Bible code as 'numerological nonsense'. They say that random distribution of letters in a text of sufficient length can generate misleading patterns much more frequently than naive intuition might suggest, and 'encoded' messages can be found in books other than the Bible. To prove this point, Australian mathematician Brendan McKay analysed a Hebrew translation of Leo Tolstoy's *War and Peace* and found, among other things, fifty-nine words related to the Jewish festival Chanukkah (Hanukkah). His analysis of the English text of *Moby Dick* revealed 'predictions' for assassinations of nine leaders, including Martin Luther King, John F. Kennedy and Indira Gandhi. The text, incidentally, also 'predicted' the death of Princess Diana. McKay pointed out that the Hebrew of the Torah contains no vowels, while English with vowels made his task much more difficult in *Moby Dick*. He assured that none of these patterns happened by other than pure random chance.

The irony of the Bible code 'predictions' is that they are 'postdictions' made after the event had happened, not before. The Danish physicist Niels Bohr, known to school students for his model of the atom, once said that prediction is very difficult, especially about the future.

Biological transmutation

13
Playing arithmetical games with atomic numbers

WHEN IN HIS youth Louis Kervran (1901–83), a French chemist, read Gustave Flaubert's satirical novel on bourgeois stupidity, Bouvard et Pécuchet (1881) he became obsessed with a reference about the formation of calcium in eggshells. He traced the original work of Louis Nicolas Vauquelin (the French chemist famous for the discovery of beryllium in 1797) in which he had wondered how the chickens he had studied could have enough calcium in their eggshells when their diet was limited to oats, which were rich in potassium but hardly contained any calcium. Vauquelin didn't provide any answer.

Years later Kervran suddenly realised that one could get to calcium (atomic number 20) from potassium (atomic number 19) when it combines with hydrogen (atomic number 1). But transmutation of elements doesn't follow such simple arithmetic.

During medieval times alchemists dreamed about the transmutation of inexpensive base metals such as lead into expensive glittering gold. It was the American-Italian nuclear physicist Enrico Fermi who set the stage for transmutation of elements when, in 1933, he showed that the nucleus of most elements would absorb a neutron, changing the atom into a new element. In most cases, this new element would be unstable and would decay into a different element. It occurred to Fermi that should be possible to prepare new elements by bombarding uranium nuclei with free neutrons (uranium, atomic number 92, is the heaviest element known to exist naturally). He tried the experiment in 1934 but was unsuccessful. In 1940 the American physicists Glen T. Seaborg and Edwin McMillan produced the first artificial element, neptunium, 93.

Alchemists' dream of transmutation was now a reality: but the process doesn't take place in alchemists' flasks. It takes place in atomic reactors and atomic bombs (where a heavy atom splits into two or more lighter atoms) or in stars (where lighter atoms join to form a new heavier atom) and is accompanied by the release of tremendous amounts of energy.

Kervran ignored all this chemistry and in 1966 published a book on biological transmutations (which was translated into English in 1972 as Biological Transmutations and Their Applications in: Chemistry, Physics, Biology, Ecology, Medicine, Nutrition, Agronomy, Geology). He notes in this book: 'In order to succeed in transmuting elements biologically it is necessary to abandon certain concepts of the so-called "exact sciences", which are exact only for simple and isolated cases foreign to biology.'

For his chicken eggshell theory, Kervran ignored the vast amounts of energy the fusion of potassium and hydrogen would release. He rationalised his theory by simply saying that this fusion took place at low energy. In another book published in 1969, he turned his attention to cold fusion in shellfish: 'A crayfish was put in a sea water basin from which calcium carbonate had been removed by precipitation; the animal made its shell, anyway.' You know Kervran's simple arithmetic used to hypothesise how calcium in shell had been fused.

For his discovery of amazing things inside living flesh and for his conclusion that the calcium in chickens' eggshells is created by a process of cold fusion, Kervran was awarded the 1993 Ig Nobel Prize for Physics. Ig Nobels, a satirical version of the traditional awards, are presented by Improbable Research, an American organisation which also publishes the magazine, Annals of Improbable Research. 'The Ig Nobel Prizes honor achievements that first make people laugh, and then make them think,' the organisers claim. 'The prizes are intended to celebrate the unusual, honor the imaginative – and spur people's interest in science, medicine, and technology.'

Cold fusion in chicken is absolutely worthy of Ig Nobel, so was cold fusion (story 18) in a test tube, which happened before Ig Nobels began in 1991. Both make us laugh and then make us think about the boundary between science and silliness.

14
Without rhyme or reason

BIORHYTHMS ARE NOT biological rhythms, which belong purely to science. Biorhythms, on the other hand, are nothing but pure bunkum.

The pseudoscientific biorhythm theory claims that we have three fixed cycles, each starting at the moment of birth: a 23-day physical cycle, 28-day emotional cycle, and a 33-day mental cycle. None of these cycles actually exist, but that doesn't stop biorhythm supporters claiming they can be used to explain human behaviour and physiology.

The original biorhythm theory was proposed in the late 19th century by Wilhelm Fliess, a German physician who was a close friend of Sigmund Freud who gave us the controversial theory of psychoanalysis (*story 75*). In the late 20th century many other biorhythm cycles have been 'discovered': a 53-day spiritual cycle, a 48-day awareness cycle, a 43-day aesthetic cycle and a 38-day intuition cycle.

You don't need any medical training to work out your biorhythms, only your date of birth, a pen and paper. Many biorhythm calculators are available on the Internet. Once you have the chart, reading it is simple: your 'good' or 'bad' days of different cycles are above or below the 'zero' line, the midpoint of each cycle. Things are likely to go wrong on critical 'critical' days, the days when cycle cross on the 'zero' line: the 'double critical' day is when two cycles cross, the 'triple critical' day when three cycles cross (you are advised not to make any important decisions on that day).

It would be nice if life were so simple. Unfortunately, it's not for scientists who study the biological clock, an internal timing system that regulates the daily rhythms of many processes in our body. Hundreds of cellular, physiological and behavioural patterns have been observed to follow a 24-hour cycle in humans. For this reason, the biological clock is also called the circadian clock (from the Latin *circa*, 'about', and *diem*, 'a day').

The circadian clock tends to be synchronised with cycles of light and dark.

Other factors such as ambient temperature, meal times, stress and exercise can also affect the clock. Jet lag and health problems associated with shift work are caused largely by the body's battle against its circadian clock when it is abruptly thrown out of phase with the light-dark cycles or sleep-wake cycles. Defective clocks can even trigger depression.

Scientists have yet to fully unravel the mysteries of the biological clock in humans, but for the supporters of biorhythms, our daily physiological and behavioural patterns are as simple as a biorhythm chart.

Bode's law

15

A law of science or a remarkable coincidence?

IN 1772 JOHANN Titius, a professor at Wittenberg in Germany, discovered a remarkable numerical relationship between the distances of the planets from the Sun. He pointed out that the numbers in the series 0, 3, 6, 12, 24, 48, 96, when added to 4 and divided by 10, produced the series 0.4, 0.7, 1, 1.6, 2.8, 5.2, 10. If Earth's distance from the Sun is set at 1 astronomical unit (AU, about 150 million kilometres or 93 million miles), then these numbers give the distances of the six planets known at the time, except for position 2.8. Titius suggested that this gap belonged to still undiscovered satellites of Mars.

That same year, the German astronomer Johann Bode picked up Titius's rule and quoted it without any acknowledgement in his astronomy textbook. However, he suggested a new planet for the gap at 2.8. The rule is now known as Bode's law. Although Bode also carried out other astronomical investigations, he is remembered today for popularising a relationship that he did not originate.

When the celebrated German-British astronomer William Herschel discovered the planet Uranus in 1781, it also fitted Bode's law (continuing the Titius series by doubling 96 for Saturn, that is, 192; when added to 4 and divided by 10, this gives 19.6, which is close enough to 19.2, the actual distance of Uranus from the Sun in astronomical units).

Astronomers now strongly felt that another planet was to be found between Mars and Jupiter, but when the 'missing planet' was discovered in 1801 by Giuseppe Piazzi, a monk and director of the Palermo Observatory in Italy, it showed many characteristics that made it quite unlike planets. However, Piazzi named the 'new planet' Ceres.

Bode was convinced that his law (which he had unceremoniously pinched from Titius) was true. He wrote to Herschel, 'I hold myself convinced that Ceres is the eighth primary planet of our solar system.' Herschel was not influenced by

Bode's plea for his law. He was sure that Ceres represented a new and different class of celestial bodies. We know now that Ceres is the largest asteroid in the belt between Mars and Jupiter.

The following table shows Bode's law as applied to planets known at the time of the formulation of the law and later discoveries. The law breaks down badly for two outer planets.

Planet	Mean distance from the Sun by Bode's law (AU)	Actual mean distance from the Sun (AU)
Mercury	0.4	0.4
Venus	0.7	0.7
Earth	1.0	1.0
Mars	1.6	1.6
(asteroid Ceres)	2.8	2.8
Jupiter	5.2	5.2
Saturn	10.0	9.5
(Uranus)	19.6	19.6
(Neptune)	38.8	30.0
(Pluto)*	77.2	39.2

* Pluto is no longer classified as a planet but as a dwarf planet

There is no generally accepted theoretical explanation for Bode's law. Proponents of the law say that it's one of the greatest mysteries of the solar system. Critics say that the so-called law is a pure coincidence. A scientific law predicts accurately; in the case of Bode's law, the deviations from prediction are far too large to be dismissed as systematic errors.

Bode's law is not necessarily pseudoscience but pseudoscience flourishes because we fail to translate algebra into everyday language where the absurdity of the argument can be easily established. The following story illustrates the point. When the Swiss-born mathematical genius Leonhard Euler became totally blind in 1766, he was invited by Catherine the Great to live in St Petersburg. Once she asked Euler to silence the visiting French philosopher Denis Diderot, who was trying to convert her courtiers to atheism. Euler said gravely to Diderot, 'Sir, $a + b^n /n = x$, hence God exists; reply.' The mathematically challenged philosopher was speechless.

A silly mathematical argument could lead to a silly conclusion – or worst, it could make you tongue-tied.

16

The real value of Palmer's 'discovery'

SPINAL MANIPULATION HAS been practised since ancient times. The fifth-century BC great Greek physician Hippocrates wrote extensively on spinal manipulation and explained how to differentiate between complete dislocations (luxations) and partial dislocations (subluxations).

In 1895 Daniel David Palmer, an American who had worked for various periods of time as a farmer, schoolmaster and grocer and dabbled in phrenology, spiritualism and magnetic healing, announced a 'discovery' that subluxated vertebrae cause 95 per cent of all disease, and other 5 per cent are caused by other displaced joints. He argued that misaligned vertebrae affect the spinal nerves, causing illness. Once the spine was adjusted, the innate energy of the body would flow freely through the nerves, allowing the body to heal itself.

During the past 100 years, chiropractic has gained widespread public acceptance, but is this acceptance based only on belief or on scientific evidence? Critics say that nothing has changed scientifically in chiropractic since its beginning. They even question that vertebral subluxation exists. Not all chiropractors now blindly subscribe to traditional subluxation theory, and there are many reformers in the profession. Yet the pseudoscientific subluxation theory is still taught in chiropractic colleges.

Most people see a chiropractor for relief of back pain. Therefore, the question of the healing value of this chiropractic treatment is important to everyone. Based on the best available evidence, Cochrane Reviews are the most respected and impartial evaluations of medical research. A 2009 Cochrane Review of a range of chiropractic treatments in 12 studies involving 2,887 participants with low-back pain (LBP) concluded: 'Combined chiropractic interventions slightly improved pain and disability in the short-term and pain in the medium-term for acute and subacute LBP. However, there is currently no evidence that supports or refutes that these interventions provide a clinically meaningful difference for pain or disability in people with LBP when compared to other interventions.' In other words, alternative chiropractic treatment is not any better than conventional mainstream treatment

Climate scepticism

17
Myths can't change the reality

When in 2007 the Intergovernmental Panel on Climate Change (IPCC) that advises the United Nations on climate change released its report on global warming it was claimed that the question mark has been removed from the debate on whether human activities are causing climate change. But the question mark became bigger when in 2009 hundreds of private emails allegedly exchanged between some of the world's leading climate scientists were stolen from the University of East Anglia's Climatic Research Unit and leaked online. The emails suggested that some scientists had manipulated the data to exaggerate the case for global warming. The question mark grew again in 2010 when the IPCC apologised for wrongly claiming in its 2007 report that there was a high chance the Himalayan glaciers would melt away by 2035.

These scandals became a new fodder for climate sceptics who question the reality of climate change. Scientists, however, claim that human activity is undoubtedly influencing climate change, even if we do not know absolute answers to 'when' and 'how much'.

Here're some of the reasons why climate sceptics think that climate change is a conspiracy theory:

Sunspots influence climate change

Sunspots are like freckles on the Sun's bright face. Wherever magnetic fields emerge from the Sun, they suppress the flow of the surrounding hot gases, creating relatively cool regions, which appear as dark patches in the Sun's shallow outer layer known as the photosphere. The number of visible sunspots varies in a regular cycle, known as the sunspot cycle, reaching a maximum about every 11 years. Near a 'solar minimum', there are only a few sunspots. During a 'solar maximum' there is a marked increase in the number of sunspots and solar flares, which are huge bursts of energy released from the region of sunspots.

Ever since the 11-year sunspot cycle was documented in 1843, scientists have been fascinated by the possibility that the cycle might influence the Earth's climate. Satellite measurements show that the solar output indeed rises and falls in synchrony with the sunspot cycle. The flickerings are weak – only 0.1 per cent change in solar energy – but they do not show any upward or downward trend. Even so, they do influence temperatures on Earth. This influence, however, is ten times smaller than the effect of greenhouse gases over the 11-year sunspot cycle.

Cosmic rays are the culprit

The Earth's atmosphere is continually being bombarded by cosmic rays. These are intense radiation that come from outside the solar system but are also generated in solar flares or other energetic solar events. Some climate sceptics claim that cosmic rays trigger cloud formation: cosmic rays ionise air which imparts an electric charge to aerosols encouraging them, in theory, to clump together to form clouds. Cloud scientists say that there is no convincing evidence to link cloud coverage to cosmic rays. Besides, measurements of cosmic rays intensity over the past 50 years show no correlation between the intensity and the global warming trend of the past few decades.

It's water vapour, not carbon dioxide

Sceptics claim water vapour accounts for about 98 per cent of all warming. Water vapour is indeed the most abundant greenhouse gas and is the big player as far as climate is concerned. However, water vapour accounts for only about 50 per cent of global warming; clouds are responsible for another 25 per cent and carbon dioxide and other greenhouses gases account for the remaining 25 per cent. Water vapour concentrations are increasing in response to rising temperatures. The dumping of carbon dioxide into the atmosphere makes the atmosphere more humid which amplifies the warming from carbon dioxide.

You can't trust computer models

Scientists' computer models cannot even predict weather accurately, how could we trust their climate models? If their models were any good, they would have saved us from global financial crisis since economists would be using them to predict the stock market. First, weather and climate are different; climate is the average weather pattern over a period of time. Second, economists also use models for financial markets, which are much trickier to model than climate. Models used by climate scientists are never perfect but they are continually tested and validated against data from various sources.

***

THE FIFTH ASSESSMENT report of the Intergovernmental Panel on Climate Change (IPCC), released in November 2014, represents the latest mainstream scientific opinion on climate change. The report says that it was *extremely likely* that the climate change since the 1950s is the product of human activity. In IPCC-speak, *extremely likely* means that the event has a probability of greater than 95%.

If an event is *virtually certain*, there is a greater than 99% probability that it will occur. In its predictions, the report also uses the terms *very likely* (greater than 90% probability) *and likely* (greater than 66% probability), *more likely than not* (greater than 50% probability) and *unlikely* (less than 33% probability). The terms used in the report for scientists' confidence in the evidence are *very high, high, medium, low* and *very low*.

Compared to the 2007 fourth assessment, the 2014 report presents stronger evidence of the ways the planet is already experiencing the effects of human-caused changes such as sea-level rise, shrinking glaciers, decreasing snow cover, warmer oceans and more frequent and intense extreme weather events.

Other trends in the report include:

- It is very likely that the number of cold days and nights has decreased and the number of warm days and nights has increased on the global scale.
- It is likely that the frequency of heat waves has increased in large parts of Europe, Asia and Australia.
- The frequency or intensity of heavy precipitation events has likely increased in North America and Europe. In other continents, confidence in changes in heavy precipitation events is at most medium.
- Global mean sea levels will continue to rise at a rate very likely to exceed

the rate of the past four decades.
- Increases in intensity and/or duration of draught: low confidence.

The report also says that greenhouses gases contributed a global mean surface warming *likely* to be in the range of 0.5 °C to 1.3 °C (0.9 °F to 2.3 °F) over the period 1951 to 2010. Further warming will continue if the emission of greenhouse gases continues.

Depending upon whether you are a climate alarmist (a term used by Fox News for those who believe in the science of climate change) or a climate sceptic, you may accept or reject the IPCC prediction.

18
Nuclear fusion in a glass jar

NUCLEAR FISSION AND fusion both release tremendous amounts of energy. In nuclear fission, the nucleus of a heavy atom breaks up into two or more lighter atoms. This reaction takes place in atomic reactors and atomic bombs. In nuclear fusion, the nuclei of light elements join to form a new, heavier nucleus. Energy in stars is produced by nuclear fusion reactions. Starting a nuclear fusion requires a temperature greater than the Sun's. Scientists have yet to build a successful fusion reactor. If a way can be found to start fusion at room temperature, it will be a potential source of limitless pollution-free and cheap energy and will solve the world's energy problems.

On 23 March 1989 Stanley Pons and Martin Fleischmann, chemists at the University of Utah, astounded the world when they announced at a press conference that they'd discovered a way to produce nuclear fusion in a glass jar at room temperature. Their bench-top fusion reactor consisted of two electrodes – one palladium, the other platinum – immersed in a glass jar of heavy water (water containing deuterium in place of ordinary hydrogen). This simple apparatus, which could be assembled in a high school lab, was claimed to produce heat energy ten times greater than the electrical energy passed through the electrodes. The chemists also claimed that the reaction generated gamma radiation. When hundreds of scientists around the world tried to replicate the experiment, the only thing they could find in their glass jars was cold water.

The cold fusion saga is one of the most extraordinary stories in the history of scientific research. It has shown that scientists are just as vulnerable to the human failings of greed and vanity as anyone else.

Most nuclear physicists believe that cold fusion is 'impossible'. However, some researchers are adamant that this energy source is real and are continuing with their own experiments. You haven't heard last of cold fusion.

Common sense and science

19

A complex relationship

'COMMON SENSE IS the collection of prejudices acquired by age eighteen.' This famous quotation is attributed to Einstein but its accuracy is questionable. What is common sense, anyway? It may be called ordinary, nonspecialised knowledge which leads someone to make a sound judgment. American cultural anthropologist Lathel F. Duffield wants to add 'common' to this meaning and calls common sense 'a judgment or opinion shared by members of a group'. He argues that this sharing makes common sense a cultural pattern that is a part of the overall cultural configuration of society. Therefore, he says, common sense is a learned way of thinking.

This learned way could also lead to prejudices, which are undesirable because they result in unfair treatment of individuals. Common sense tells us that the Sun moves around Earth. In the 16th century when the Polish astronomer Copernicus rejected this commonly held wisdom, he was ridiculed as a fool who wanted to turn the whole science of astronomy upside down. Since then, common sense has evolved and now anyone who believes that that Earth is the centre of the solar system would be ridiculed as a fool (*story 36*). When common sense fails to evolve, it becomes ignorance, prejudice, even bigotry. We should not confuse common sense with prejudice.

The advent of quantum mechanics in the early 20th century introduced phenomena that were beyond the capabilities of our senses (even Einstein was 'spooked' by quantum entanglement, (*story 89*). The hardware and software 'wired' in humans by evolution had reached its peak. Science couldn't be simply guided by common sense. 'A classic example of the limitations of our neural wiring is the inability to picture more than three dimensions,' says American theoretical physicist Leonard Susskind. 'Physicists have had no choice but to rewire themselves. Where institution and common sense failed, they had to create new forms of intuition, mainly through use of abstract mathematics.' Still, we must use this uncommon sense sensibly, he advises.

As science evolves so does our common sense and they define each other as they evolve together. Remarks the late British theoretical physicist John Ziman in *Real Science: What It Is, and What It Means* (2000): 'Thus, when we contrast a scientific belief with "common sense" we are indicating that we think we could define it precisely and give some coherent account of why it should be relied on, rather than just "taking it for granted" as an undeniable truth.'

Now, if proponents of intelligent design (*story 42*) move away from the common-sense belief in human's uniqueness in the world (mainly based on religion) and try to appreciate scientific belief of proponents of evolution, we may find a common ground. This will be the new common sense and the new science of creation–evolution.

Consciousness

20

Explaining the unexplainable?

UNTIL RECENTLY NEUROSCIENTISTS ignored the study of consciousness: the problem was considered either philosophical or too difficult to study experimentally. But most now believe that consciousness – our immediate, subjective awareness of the world and ourselves – is likely to be explainable as the behaviour of the brain's 100 billion neurons.

The human brain is gradually yielding its secrets to increasingly powerful tools that can image what happens inside a person's head, say, when listening to a song. These brain-imaging tools range from PET (positron emission tomography, which uses radioactivity to label neurotransmitters), fMRI (functional magnetic resonance imaging, which uses a powerful magnetic field to align atomic nuclei in the brain) to MEG (magneto-encephalography, which can pick up the faint magnetic fields generated by active nerve networks). Neuroscientists can even 'film' the trails of neural activity blazed by flitting thoughts and feeling. One day these 'films' could even reveal the nature of consciousness itself.

Francis Crick, who shared a Nobel Prize with James Watson for the discovery of DNA's structure in 1953, should receive much of the credit for the current scientific interest in looking for an explanation of how processes in the brain creates consciousness awareness. In 1990 he and Christof Koch, a young neuroscientist who collaborated closely with Crick, rejected the belief of many of their colleagues that consciousness cannot be defined, let alone studied. Consciousness is a legitimate subject for science, they declared. Crick died in 2004 but Koch is still actively involved in research on consciousness.

According to Crick and Koch, one cannot hope to achieve a true understanding of consciousness by treating the brain as a black box (an object whose internal structure is irrelevant). Only by examining neurons and the interactions between them could scientists accumulate the kind of knowledge required to create a scientific model of consciousness. Crick and Koch believe that, though there are many possible approaches to the problem of consciousness, they have focused on 'visual awareness rather than other forms of consciousness, such as pain or self-awareness, because humans are very visual animals and our visual awareness is especially vivid and rich in information.'

David J. Chalmers, an Australian philosopher, believes that philosophy must bridge the 'explanatory gap' between a physical theory of consciousness and our subjective experience. 'I got into this field to try and understand the problem of how a physical system like a brain could also be a conscious being with subjective experience,' he says.

Chalmers makes a distinction between the 'easy problems' and the 'hard problem' of consciousness. His list of easy problems includes: How is it that a brain can discriminate information from the world? How is it that it can bring it together in the brain and integrate it? How is it that the brain or a human being can verbally report their mental states? How is it that we bring information to bear in controlling our action?

Chalmers does not consider the easy problems as trivial problems, but he believes that continued work in cognitive psychology and neuroscience will answer them. But the 'hard problem' would still remain: how physical processes in the brain give rise to subjective experience. 'Why is that physical processing in the brain, no matter how sophisticated, should give rise to any subjective inner life at all, why couldn't that have all gone in the dark?' he says, 'That's the real mystery.'

To illustrate the distinction between the 'soft problems' and the 'hard problem' Chalmers uses a thought experiment devised by the Australian philosopher Frank Jackson: Mary, a neuroscientist in the 23rd century who knows everything there is to know about how the brain processes colour, has lived all her whole life in a black-and-white room. Mary does not know what it is like to see a colour such as red. 'It follows that there are facts about conscious experience that cannot be deduced from physical facts about the functioning of the brain,' says Chalmers.

Daniel C. Dennett, an American philosopher, is the leading critic of the 'hard problem'. His 'multiple drafts' theory says that consciousness is not a unitary process but rather a disturbed one. The brain is a kind of hypothesis-making machine, constantly throwing up new 'drafts' of what is going on in the world. The sequential timing of events breaks down at extremely small time scales within the brain, and the events that make up consciousness cannot be ordered. There is no central place in the brain where everything is presented or decisions are made. 'Mental states do not become conscious by entering some special chamber in the brain,' he stresses.

How the processes of the brain translate to consciousness is still a mystery to neuroscientists. Even if they unravel this mystery, will it really explain consciousness? There are many who still hold to the view that our minds are more than the brain and therefore consciousness will remain the ultimate mystery. They are uncomfortable with the idea of reducing their minds to pieces of meat and their private ideas and feelings to pixels on computer screens.

Perhaps clues to the mystery of consciousness lie in old definitions of a scientist and a philosopher. A scientist looks for a black cat in a dark room while a philosopher looks for a black cat in a dark room where there is no black cat. Where is the black cat called consciousness?

Cosmic collisions

21
Planets colliding on the edge of science

IN ABOUT 1500 BC a comet was ejected from Jupiter. It came close to Earth and Mars before settling into an orbit between Earth and Mercury and becoming the planet Venus. As Earth passed through the comet's tail, it caused a range of events: meteoric bombardment started widespread fires; falling meteoric dust turned rivers blood red and plunged the world into darkness that lasted four days; Middle East oil fields were created as large quantities of petroleum fell from the sky, and edible carbohydrates (manna) fell on Earth. For hundreds of years, multiple collisions occurred between Earth, Venus and Mars, which ended in 700 BC. Earth suffered extreme geological changes from these close passes of Venus and Mars, causing catastrophic events such as volcanic eruptions, earthquakes, global floods and the formation of new mountain ranges. The evidence for these cosmic collisions can be found in myths and legends of all cultures; for example, in Biblical stories such as Noah's flood, parting of Red Sea, Joshua making the Sun stand still, and people eating manna from the sky.

All these claims of violent catastrophes and many more appeared in a book, *Worlds in Collision*, published in 1950. Its author, Immanuel Velikovsky, was a Russian-American psychiatrist who died in 1979. The book unleashed a torrent of arcane information. Apparently, it impressed book critics and they gave it highly favourable reviews in popular magazines and newspapers, which made the book hugely popular (it's still in print). Astronomers were not so impressed with this cataclysmic concept of world history.

All of Velikovsky's theories have been thoroughly rejected, but there are still many dedicated followers. Writes David Morrison, an American space scientist, in 'Velikovsky at Fifty', an article published in 2001 in the *Sceptic* magazine: 'There is ample evidence that Velikovsky was little more than a crank, something that was evident to the astronomers from even a cursory look at his book.' Reflects Brian Toon, an American atmospheric scientist: 'Velikovsky influenced me by showing how public is so easily fooled by pseudoscience.'

Crops circles

22
Reaping by humans, aliens or nature?

THE TERM 'CROP circles' refers to huge geometrical patterns that appear somewhat mysteriously, usually overnight, in fields of wheat, barley, rye, oats and other cereal crops. The plants are rarely cut or damaged but their stalks are swirled and flattened.

Crop circles have been reported since the 1960s but the phenomenon attracted worldwide popular media attention in the northern summer of 1980 when several circles, each about 18 metres (60 feet) in diameter, appeared randomly in a field of oats in the rolling Wiltshire countryside in southern England. Since then thousands of circles have been reported; some in other countries but mostly in southern England. As their numbers have increased so has the complexity and size of their designs. Relatively simple circles of the 1960s have evolved into elaborate pictograms and shapes (sunflower, scorpion, spider web, the Buddhist wheel of Dharma, the Pythagorean symbol of wellbeing) and complex mathematical and fractal patterns (ten digits of pi, Euclid's theorems, Ptolemy's theorem of chords, a vortex of logarithmic curves). A formation that appeared in 2001 (in Wiltshire, of course) covered about five hectares (twelve acres) and was more than 243 metres (800 feet) across.

Most of the recent crop circles are the work of hoaxers prompted by media attention. But those who study crop circles (they call themselves 'cerealogists') reject this mundane idea and offer other exotic explanations: (1) they are the work of intelligent aliens as most crop circles appears close to UFO sightings; or (2) they are caused by spiritual energy or nature spirits. There is a conspiracy theory as well: they are formed by the testing of some kind of secret microwave weapon the government has developed and do not want to tell the public about it.

Scientists, however, offer simple explanations. Exactly a century before the appearance of the famous Wilshire circles, an amateur English scientist J. Rand Capron described strange circular spots in a wheat field in a letter to the journal Nature of 29 July 1880, 'Examined more closely, these all presented much the same character, viz., a few standing stalks as a centre, some prostrate stalks with their heads arranged pretty evenly in a direction forming a circle round the centre, and outside these a circular wall of stalks which had not suffered ... and I could not trace locally any circumstances accounting for the peculiar forms of the patches in the field, nor indicating whether it was wind or rain, or both combined.' He concluded that the phenomenon was probably caused by 'some cyclonic wind action'.

Modern scientists also suggest that they are formed by an atmospheric vortex, some kind of swirling air current similar to a tornado. Others postulate that this vortex must be electrified for it to make such complex patterns. Critics say that even an electrified vortex cannot cut such precise patterns on the ground.

If all crop circles are not manmade, then scientists have a challenging task ahead of them. The problem is that most 'credible' scientists do not like to involve themselves in weird and off-beat things.

THE FOLLOWING 1687 woodcut pamphlet entitled 'Mowing-Devil' tells a tale about an English farmer who asked the village mower to cut down his crop of oats. When the mower demanded too much money, the farmer replied that the devil should mow it rather than you. The whole field was cut down mysteriously the same night. The illustration shows the devil with a scythe mowing in a circular design, not just a small section as in the case of modern crop circles.

The Mowing-Devil:

Or, Strange NEWS out of Hartford-shire.

Being a True Relation of a Farmer, who Bargaining with a Poor *Mower*, about the Cutting down Three Half Acres of Oats, upon the *Mower's* asking too much, the Farmer swore, That the Devil should Mow it, rather than He. And so it fell out, that that very Night, the Crop of Oat shew'd as if it had been all of a Flame, but next Morning appear'd so neatly Mow'd by the Devil, or some Infernal Spirit, that no Mortal Man was able to do the like.

Also, How the said Oats ly now in the Field, and the Owner has not Power to fetch them away.

Licensed, August 22th, 1678.

Cryptozoology

23

Searching for snarks

CRYPTOZOOLOGY TRIES TO cloak monsters of myth, fantasy and nightmare in scientific impartiality. The word was coined in the late 1950s by Belgian zoologist Bernard Heuvelmans and literally means 'the study of hidden animals'. Cryptids, or 'hidden animals', are unknown or supposedly extinct animals; and 'zoologists' studying them are 'the snark set' (a moniker bestowed upon them by a commentator in the London *Times* of 20 August 1984). Snark is the nonsense word coined in 1876 by legendary Lewis Carroll in his book, *The Hunting of the Snark*. It now means an imaginary animal.

The list of modern snarks cryptozoologists study is long; it has more than 250 cryptids from around the world. The famous ones are: Thylacine or the Tasmanian tiger which is believed to have become extinct in the last century; Bigfoot, a hairy gorilla-like 2.5 metre (8 feet) tall biped primate that inhabits forests in North America and its cousin Yeti of the Himalayas; a 30-metre (100-foot) snake known as the giant serpent of South America, Ogopogo that inhabits Lake Okanagan in British Columbia and its famous cousin Loch Ness monster (*story 48*), the star of countless books, documentaries and movies.

It's true that some cryptids have migrated from mythology to biology textbooks; for example, the gorilla in 1847, the giant panda in 1869, the short-necked giraffe in 1901; the pygmy hippopotamus in 1913; the bonobo or pygmy chimpanzee in 1929; and the giant gecko in 1984. This doesn't imply that every mythical cryptid would be discovered some day.

Cryptozoologists are good at dragging monsters from mythology or native lore and supplying grainy photos or murky videos of the monster as an evidence. But they never ever produce the body. Science demands tangible evidence. Would it not be more beneficial to science if cryptozoologists spent their energies on studying some of the species of organisms that are disappearing at a faster rate than they can be classified, sparing snarks for children's books?

Crystal healing

24

It's all in the mind

WHEN WE THINK of crystals, we think of glasslike solids with regular symmetrical shapes. Some crystals such as diamonds, rubies, emeralds and other precious gemstones have a charming and often calming appearance. These aesthetic properties have made them attractive to almost all civilisations throughout the ages. Priests, shamans and wizards have associated crystals, the stones thrown down from the heavens by the gods, with magical healing and psychic powers.

No wonder New Age healers have dusted off old 'crystal powers' and assigned them an important role in their esoteric beliefs that are totally out of sync with science. They claim that crystals, especially quartz, can heal ailments and diseases when worn regularly. Not only this, quartz crystals can improve and amplify psychic powers when properly charged with positive energies. Charging requires first cleaning the crystal in sea water or cold running water and then clearing it by holding it in your hand and concentrating on it. Once the crystal is clear, it can be charged by concentrating on the thought pattern with which you want it to be programmed. For example, you may say loudly or silently think: 'May only the purest, highest and finest energies be channelled through this crystal and it may work always in accordance with divine will.' The crystal receives and locks the vibrations of your thoughts.

Quartz is the most common mineral in Earth's crust; chemically it is silica (silicon dioxide). Quartz and some other crystals show piezoelectric effect; that is, when the crystal is compressed or stretched it produces an electric voltage. Alternatively, when an electric voltage is applied to it, it's either compressed or stretched. In this way, the crystal responds to an alternating voltage in a periodic movement. The frequency of a quartz crystal depends on the dimensions of its shape or size, so it can control timing in a quartz device. The piezoelectric effect was discovered in 1880 and piezoelectric devices are now used widely in the industry.

The New Age belief in the healing and psychic power of crystals such as quartz is based on a misunderstanding of the piezoelectric effect. While quartz vibrates at millions of cycles per second, it cannot affect brain waves which have markedly lower frequencies. There is absolutely no experimental evidence that quartz can either alter brain waves or brain waves can alter quartz frequencies.

Psychologists say that the power of the crystal is in the mind rather than in the crystal itself. Our expectations and beliefs can greatly change the course of an illness (*story 70*). If you're still not convinced, heed the advice that even a cheap quartz crystal would achieve the same alleged healing effect than an expensive crystal pendant or bracelet. Why waste money?

Delphic oracle

25

Inhaling vapours and prophesising

THE TEMPLE OF Apollo at Delphi on the southern slopes of Mount Parnassos was the most popular religious site in the ancient Greek world. First built in the seventh century BC and rebuilt many times over the centuries, the temple's foundations survive today along with several Doric limestone columns.

Apollo was the god of prophecy and inside the temple sat the Delphic oracle who responded to questions about the future from rulers, philosophers and citizens. The Pythia, the high priestess who presided over the oracle, was the medium through which Apollo spoke. Pythia's role was filled by different priestesses for about a thousand years until the temple was closed in the fourth century AD. The oracle's prophecies figure prominently in Greek myths. When Oedipus went to Delphi to ask about his true parents, the oracle warned him that he would murder his father and marry his mother.

Pythia sat on a tripod in a small cavernous chamber holding a sprig of laurel (Apollo's sacred tree) in one hand and in the other a cup containing water from a spring beneath the temple. During oracular sessions, she appeared as if she was in a trance and spoke in an altered voice. Her utterances would then be interpreted by the attending priest in response to the seeker's question. Many ancient writers, including Plutarch, the famous first-century essayist and biographer who served at Delhi, have attributed the oracle's prophetic powers to fumes rising out of a fissure in the ground or from the spring. Plutarch said that the air in the chamber smelled sweet like flowers.

Did the intoxicating vapours from the ground loosen the lips of the Pythia? In 1927 French geologists surveyed the temple and found no evidence of fissures or rising gases. They dismissed the ancient explanations as myths.

A recent geological study has shown that Plutarch was indeed right. Greece sits at the confluence of three tectonic plates and there is a major fault running east to west directly under the temple. A second fault runs north to south and interacts with the east-west fault below the temple. The limestone rock below the temple is also full of cracks and able to pass groundwater and gases. An analysis of gases in modern spring water showed that it contained measurable amounts of methane and ethane. These gases mix with groundwater and emerge around springs.

The geology of the temple site suggests that most likely it also produced ethylene. Ethylene – a sweet-smelling gas once used as anaesthetic – could have arisen through fissures created by faults. In light doses, ethylene produces a mild trancelike state, just what the oracle might have experienced.

26

What really killed our lovable leapin' lizards?

SOME 65 MILLION years ago a large asteroid plunged out of the sky and hit Earth, throwing up a great cloud of dust that quickly covered the planet like a blanket, blocking sunlight for several years. This extraterrestrial impact wiped out the dinosaurs, along with nearly 75 per cent of all other species. This is the asteroid theory of the demise of the dinosaurs and it's now widely accepted. But it doesn't imply that the matter has been concluded forever. In the past, there have been more than one hundred attempts to explain the death of the dinosaurs, and there would be further attempts in the future. Here are three imaginative theories that have remained on the borderlands of science.

Doped

Flowering plants, the angiosperms, evolved around the same time as the dinosaurs died. Many of these plants contain poisonous substances. Modern animals avoid them today because of their bitter taste. Ronald Siegel, an American psychopharmacologist, has suggested that dinosaurs had neither the taste for the bitterness nor livers effective enough to detoxify the substance. They died of massive overdoses.

British palaeontologist Anthony Hallam has looked at the emerging angiosperms from a different angle: the dinosaurs died because of constipation caused by eating the flowering plants that replaced ferns, a dinosaur dietary staple containing laxative oils.

These lateral thinking exercises – poison and constipation – have been knocked on the head by those who say that the angiosperms appeared 40 million years before the dinosaurs' death.

Victims of cancer

Don't blame drugs – it was cancer. This novel but serious explanation comes from American astrophysicist Juan Collar. He claims that dinosaurs were wiped out by epidemics of cancer. No, it wasn't caused by smoking. The cancer was triggered by massive bursts of neutrinos released by dying stars. In the final stages of their death, massive stars radiate most of their energy in the form of neutrinos. These dying stars are not as bright as supernovas, and therefore difficult to find.

Collar calls them 'silent' dying stars. He predicts that a silent star death occurs within 20 light years of Earth about once every 100 million years. He suggests that a collapsing star would produce twelve malignant cells per kilogram of tissue, each of which could trigger a tumour. The effect would be more severe in dinosaurs because they had more tissue to become cancerous. He advises cataloguing possible 'neutrino bombs' – sources of neutrinos – in the galaxy to save us from the same fate as that of the dinosaurs.

Hot weather equals infertility

American palaeontologist Dewey McLean has suggested that the dinosaurs died because of a slight but critical increase in the global temperature. The effect of the heat was not to actually kill the dinosaurs but effectively to castrate them. Because large animals do not shed excess heat as efficiently as small animals do, a temperature increase of just 2 degrees could have baked the considerable reproductive apparatus of a 10-tonne male dinosaur enough to kill its sperm.

The argument that McLean presented to support his theory went something like this. Numerous completely unhatched dinosaur eggs have been found in rocks, possibly suggesting a failure of fertilisation. At the same time, eggs show thinning of the shells. Modern birds under stress also lay thin eggs. Put two and two together and you have a theory: hot weather stressed dinosaurs; stress made them infertile.

27

Does it work?

DOWSING IS THE art – some claim it to be the science – of finding underground water, petroleum, minerals, buried treasures, enemy booby traps and even lost articles and people. Its most widely known application is looking for underground water, when it is known as water divining or witching.

Dowsers use divining rods which are usually Y-shaped twigs, sticks or rods – customary woods are hazel, peach or willow, but others can also be used. Map dowsers use a pendulum over maps. However, archetypical dowsers are the ones walking around with forked rods looking for underground water. They grip the forks firmly in each hand, the stem pointing upwards. As they walk over the area they are surveying, suddenly the rod would twist violently in their hands. If the rod points downwards, it indicates where to dig for water. Some dowsers use two rods; the rods cross when they are above water.

Dowsing is as old as pyramids of Egypt. Ancient Egyptians and Greeks used it for predicting future events. In Europe in the Middle Ages it was called the work of devil and the Church frowned upon the practice. It gained respectability in the 18th century and then people saw in it the hand of God. Sceptics still frown upon the practice, but they agree that the rod serves only as a vehicle for signalling an effect produced by the dowser. Dowsers believe that there are 'vibrations' from all objects and their subconscious mind detects them. George Applegate, an English engineer with long dowsing experience, says that the subconscious mind is not an entity in itself; it is part of a universal subconscious mind, used and partly controlled by the individual. 'All dowsing power comes from within, and is therefore under own control,' he writes in his book, *The Complete Guide to Dowsing* (1997). 'We can direct our thought processes, consciously applying them to any circumstances.' Is it the power of suggestion that causes the rod to move again and again?

To engineers and scientists who make a living in the groundwater industry, dowsing for water is akin to astrology. Appealing but unsupported by scientific evidence. All dowsing studies done under controlled conditions have shown that dowsers do no better than chance in finding water.

Geologists assure us that underground water is so widely distributed that in most settled parts of the world, it is not possible to dig a deep hole or drill a well without finding underground water. Psychologists explain dowsing rods movements as an ideomotor effect, an unconscious body movement made in response to an idea. It is the dowser's expectations that cause the unconscious movement of the dowsing rod.

Drake's equation

28
Sheer speculation

AMERICAN ASTRONOMER FRANK Drake is a pioneering SETI (search for extraterrestrial intelligence) researcher. In 1960, he became the first person in history to use a radio telescope to listen to ETs. In those early days of SETI, many scientists ridiculed the idea of extraterrestrial intelligent life, but to Drake the idea of other intelligent civilisations beyond Earth was a distinct possibility.

In 1961 Drake invited a dozen scientists to the first-ever SETI conference. While he was preparing for the conference he came up with an equation consisting of astronomical, environmental, biological and cultural parameters to estimate the number (N) of advanced civilisations that exist now in our galaxy and are capable of communicating across interstellar distances:

$$N = R_* \times f_p \times n_e \times f_l \times f_i \times f_c \times L$$

where
R_* = number of stars born each year in our galaxy (astronomical)
f_p = fraction of these stars that have habitable planets (astronomical)
n_e = average number of planets or moons suitable for life around each star (environmental)
f_l = fraction of worlds on which life actually appears (biological)
f_i = fraction of worlds on which life evolves to an intelligent form (biological)
f_c = fraction of worlds on which the intelligent life can communicate to other worlds (cultural)
L = average lifetime of such technological civilisations (cultural)

This equation is now known as the Drake equation. 'It amazes me to this day to see it displayed prominently in most textbooks on astronomy, often in a big, important-looking box. I've seen it printed in *The New York Times*,' writes Drake in his book, *Is Anyone Out There?* (1991), co-authored with Dava Sobel. Since then the equation has become much more popular and ubiquitous.

But can it provide an accurate estimate? Some scientists call it speculative in the extreme. Jill Tarter, a prominent SETI researcher (the character of Dr Ellie Arroway in Sagan's book and movie *Contact* was probably based on Tarter), says that it is 'really nothing more than a systematic way of quantifying our ignorance'.

Nevertheless, if we assign values to the seven factors on the right-hand side of the equation, we can calculate N. From the current state of research in astronomy, we can assign the following approximate values:

R_* = 10 (Of all the seven factors, this is the only one supported by observational evidence beyond the solar system. We know now that there are about 250 million stars in our roughly 13-billion-year-old universe. Therefore, the average rate of star-birth is about twenty stars per year. But the figure of 250 billion stars includes all stars of all ages from 0 to 13 billion years and does not include stars from the early days of the galaxy which have already died.)

f_p = 0.2 to 0.9 (Drake estimated this value to be about 0.5.)

n_e = 0.1 to 1 (Using our solar system as a guide, Drake estimated between 1 and 5 planets.)

f_l = 0.33 (Some scientists argue that this value is 1 because life is virtually inevitable on any habitable world; others give it a pessimistic value of one in a million, as they believe that the chances of life appearing on a planet or moon are extremely small.)

f_i = 0.05 to 1 (Based on the pessimistic view that it took a very long period of time for intelligent life to evolve on Earth to the optimistic view that once primitive life appears, it will surely evolve into an intelligent form.)

f_c = 1 (This value is taken on the simple reasoning that once you have intelligent life like ours, it's likely to develop capabilities to communicate at interstellar distances.)

L = 300 to 10,000 years (This is the most uncertain value as don't have any basis for a reliable estimate – even the only known technological civilisation has been communicating with radio waves for only about 100 years. Will our civilisation last a million years or will we destroy ourselves in the near future?)

Now, by multiplying the seven terms we can arrive at two values for N = 1 or 30,000

These numbers show only what we already know: either we are alone or there are numerous worlds with intelligent life that have the capability to communicate with us. The Drake equation is simply a mathematical way of saying 'we don't know'.

Even if there are intelligent beings out there, would they develop technology which they would then deploy to send messages or meet other intelligent beings? Human psychology is no guide that they would be as interested as we are in looking for intelligent neighbours in our galaxy.

29
Power line paranoia and mobile phone mania

EVERY ELECTRICAL OR electronic device produces some form of the electromagnetic field. Radio waves, microwaves, visible light, infrared and ultraviolet radiation, X-rays and gamma rays all are forms of electromagnetic fields and are part of the electromagnetic spectrum. They all have the same speed as light (in empty space; much slower when passing through something like a wire) and behave the same as light. They differ from each other from their frequency, which is measured in hertz (cycles per second). The frequency range is vast and ranges from 30 hertz to more than 300 etahertz (one etahertz is a very large number, 1 followed by 18 zeros).

Electromagnetic waves are broadly classified as ionising and nonionising radiation. The waves consist of small particles called photons and the energy of a photon increases as the frequency of wave increases. Photons of ultraviolet light, X-rays and gamma rays carry enough energy to remove an electron from an atom or molecule. This process is known as ionisation. Ultraviolet light, X-rays and gamma rays are ionising radiation and because of their high energies are able to penetrate living cells and damage DNA. It is, therefore, advisable to limit exposure to the sources of ionising radiation.

Photons of radio waves, microwaves, visible light and infrared radiation have low energies and cannot damage atoms or molecules. Everyone is exposed to this nonionising radiation and exposure to some forms of this radiation is increasing significantly as technology advances. Health hazards of nonionising radiation are small, but medical experts warn that exposure to high levels of radio waves can damage the nervous system. Microwaves with frequencies below 3,000 megahertz can penetrate outer layers of skin and can result in burns, cataracts and possibly death.

In recent years, news headlines have highlighted the 'link' between cancer and either use of mobile phones or residing near high-voltage power lines. Let's us look at the facts.

Power lines

Both electric and magnetic fields exist close to lines that carry electricity and close to appliances. Electric and magnetic fields in homes and offices depend upon the distance from power lines, the configuration and position of electrical wiring and the number and types of electrical appliances in use. Electricity is transmitted at a frequency of either 50 or 60 hertz, and electromagnetic fields associated with these frequencies (and up to 300 hertz) are classified as extremely low frequency (ELF) fields. The photon energies at this level are too low to interact with atoms or molecules in any meaningful way.

According to the World Health Organisation, 'there is no convincing evidence that exposure to ELF fields causes direct damage to biological molecules, including DNA'. There is no need to panic about health effects of power lines. More dangerous to your health is how you view high-voltage lines: they have the power to spoil the view and stress you psychologically.

Mobile phones

No electronic gadget is as ubiquitous as the mobile phone: there are nearly five billion mobile phones in use in the world. Mobile phones emissions range in frequency from about 450 to 2,700 megahertz (electromagnetic waves below 1,000 megahertz are radio waves; those above are microwaves). Unlike microwave ovens, their peak powers are in the range of only 0.1 to 2 watts, too weak to cook human tissue. Besides, a mobile phone transmits power only when it is turned on, and the power or frequency decreases rapidly with increasing distance from the phone.

According to the World Health Organisation, 'a large number of studies have been performed over the last two decades to assess whether mobile phones pose a potential health risk. To date, no adverse effects have been established for mobile phone use.' However, the organisation warns, 'the research has shown an increased risk of road traffic injuries when drivers use mobile phones (either handheld or "hands-free") while driving.'

Studies are still undergoing to assess potential long-term effects of mobile phone use. Even if they exist, they are likely to be minuscule. So worrying about mobile phones would simply cause psychological stress, which has more potential to damage DNA and cause cancer than your mobile phone itself.

If you are still worried, experts suggest to limit the use of mobile phones or, at least, make sure the antenna, which is the largest source of radiation, is further away from your head.

Energy from antimatter

30
Propelling fictional starships

IN THE EARLY 20th century when the electron and the proton were the only known fundamental particles, physicists often wondered why electrons were always negatively charged and protons positively charged when the laws of physics were quite symmetrical with respect to charge. In 1928 the gifted British theoretical physicist Paul Dirac predicted that the electron should have a positively charged counterpart: 'This would be a new kind of particle, unknown to experimental physics, having the same mass and opposite charge as the electron. We may call such a particle antielectron.'

The symmetry between positive and negative charges in Dirac's theory also demanded an antiproton. At first, scientists were sceptical about the idea of antielectrons and antiprotons, but the discovery in 1932 of the antielectron (now known as the 'positron', short for 'positively charged electron') in the cosmic radiation by the American physicist Carl Anderson vindicated Dirac's bold prediction. Twenty-three years later, scientists at the University of California at Berkeley created the antiproton in a particle accelerator. We now know that every fundamental particle has an antiparticle – a mirror twin with the same mass but opposite charge. The idea of antiparticles is now also applied to atoms – antiatoms, which make up the antimatter.

When antimatter meets ordinary matter, they annihilate each other and disappear in a violent explosion in which mass is converted into energy as dictated by Einstein's famous equation $E = mc^2$, where E is energy, m is mass and c is the speed of light. The energy released in matter–antimatter annihilation is awesome: in a collision of protons and antiprotons, the energy per particle is close to 200 times that available in a hydrogen bomb.

If matter and antimatter annihilate each other, there is no likelihood of antimatter existing on Earth, or even in the solar system. The solar wind, the spray of charged particles emitted by the Sun in all directions, would annihilate antimatter. However, scientists speculate that antimatter could exist in distant parts of the universe, but so far they have found no evidence. This has not stopped them from creating antimatter in the laboratory.

A team of scientists at CERN, the European particle physics lab in Geneva, did just that in early 1996. For about 15 hours they fired a jet of xenon atoms across an anti-proton beam. Collisions between anti-protons and xenon nuclei produced electrons and positrons. These positrons then combined with other anti-protons in the beam to make anti-hydrogen, the simplest antiatom. Scientists could detect nine anti-hydrogen atoms. Hydrogen is the simplest (just one electron orbiting a single nuclear proton) and most abundant (it makes up about 75 per cent of the universe) of 114 chemical elements known to us. An antihydrogen atom would have a positron orbiting a single antiproton. Since 1996, CERN scientists have been regularly synthesising antihydrogen atoms, but these antiatoms exist only for about 40 billionths of second before destroying themselves by colliding with particles of ordinary matter.

So now there is the experimental proof that antimatter does exist, what can it be used for? Because the annihilation of matter and antimatter creates enormous amounts of it, it is tempting to look at antimatter as a potential source of energy. This energy might one day provide the fuel for interstellar voyages, the same way matter–antimatter annihilation powers Star Trek's fictional spaceship *Enterprise*. The amount of antimatter required for space flights is unbelievably small. A few hundred micrograms could fuel a spacecraft to Jupiter, and the round trip would take only a year. But there is a problem. The current means of producing antiparticles requires far greater energy than that generated by matter–antimatter reactions.

If you find the idea of producing energy from antimatter a bit too far-fetched, then what about the idea of an antiuniverse – a universe parallel to ours. Enter it and you will find your antimatter counterpart: anti-you. Don't shake hands – you'll annihilate each other. Would it be safe to enter into a mirror-matter world (*story 54*)?

31
Can evolution explain how we think and behave?

AT THE END of his magnum opus, *On the Origin of Species* (1859), Charles Darwin declared that 'in the distant future ... psychology will be based on new foundation.' Little he knew that one day psychologists would draft his theory of evolution to argue that everything from children's alleged dislike of spinach to the crime of rape derives from our Stone Age ancestors.

In the 1990s, more than 140 years after the publication of the great book, evolutionary psychology finally emerged as a full-fledged discipline. This new and controversial 'science' claims to explain all aspects of our behaviour, emotions and beliefs in terms of biological evolutionary processes. If our bodies evolved why not our minds? Critics, however, say in books such as *Alas, Poor Darwin: Arguments Against Evolutionary Psychology* (2000), that the grand claims of popular psychology rest on shaky empirical evidence and flawed premises. In the hands of some evolutionary psychologists Darwin has replaced Karl Marx and Sigmund Freud as the great interpreter of human existence, they say.

Let's look at some claims about how the human mind evolved. In *Scientific American* (January 2009), David J. Buller, professor of philosophy at Northern Illinois University, points out four fallacies of what he calls 'pop evolutionary psychology':

First, analysis of adaptive problems faced by our Stone Age ancestors, such as how to compete for food and mates, gives clues to the mind's design. Buller rejects this notion by saying that as we don't have knowledge of our ancestors' psychological traits, we cannot be sure about how evolution changed these traits to create the minds we have now.

Second, we know or can discover how language evolved. Buller's answer: we need to understand the adaptive functions language served among early humans, for which we have little evidence.

Third, modern humans harbour Stone Age minds. Buller replies that humans have changed physiologically since the Stone Age, so have their minds.

Fourth, data gathered by evolutionary psychologists provides clear evidence for claims such as differences in the basis of jealousy in males and females. Buller dismisses the evidence as inconclusive and says that both sexes may possess the same mechanism but they respond differently when faced with different types of threat to a relationship.

Poor old Darwin (incidentally, he left a legacy worth millions of dollars in today's money and lies buried honourably in the Westminster Abbey) gathered a great mass of evidence before he arrived at his revolutionary theory. Some human behaviour may have a Darwinian basis, but evolutionary psychology remains a shoddy science until it meets the exact standards of evidence required in science.

32
Perception beyond belief

DAN BROWN'S MEGASELLING novel *The Lost Symbol* (2009) made the word 'noetic' popular; the word comes from the Greek word *nous*, inner knowledge. In the early 1970s, the American astronaut Edgar Mitchell coined the term 'noetic sciences' and founded the Institute of Noetic Sciences to study paranormal phenomena. While on the Apollo 14 moon mission Mitchell is believed to have conducted telepathy experiments with friends on Earth.

Bertrand Russell, the English philosopher and mathematician who won the 1950 Nobel Prize for Literature, once said, 'Man is a credulous animal, and must believe something; in the absence of good grounds for belief, he will be satisfied with bad ones.'

He would have certainly categorised paranormal beliefs as 'bad ones'. The paranormal is beyond normal; the word comes from the Greek prefix *para* meaning 'beyond'. Paranormal powers are normally divided into two types: extrasensory perception (ESP) and psychokinesis (telekinesis). Parapsychologists who study these alleged phenomena use the word 'psi' (from the Greek *psyche*, mind) to refer to both ESP and psychokinesis; and psychics are people who are said to possess powers of psi.

How can one become a psychic? Some people believe they were born with psychic powers. Some claim that they gained these powers after an accident or a traumatic event. Others try to learn them.

Those who are not psychic experience the world through five senses: vision, sound, touch, taste and smell. Our normal sensory perception is limited by the limitations of our senses. For example, our sense of sound is limited to a frequency range between 20 and 20,000 hertz (cycles per second), and our vision is limited to wavelengths from about 400 to 700 nanometres.

ESP has no such limitations, as it is supposed to be gained without the use of these five senses. This extra sense can be the ability: (1) of two minds to communicate through an unknown channel (telepathy); (2) to be aware of an unknown object or event (clairvoyance); (3) to know future events (precognition); or (4) to know past events (retrocognition).

After decades of research, parapsychologists cannot even explain how ESP works. As these days many snake-oil theorists look for evidence in quantum mechanics, it's not surprising that some parapsychologists try to relate ESP to quantum nonlocality, an idea from quantum mechanics. Imagine this book is sitting on a table in front of you. To move it you have to touch it. We can only affect objects we can touch. We say we experience the world as local. Newton's discovery of gravity introduced the idea of action at a distance or what we call classical nonlocality. In the quantum world nonlocality or action at a distance comes from quantum entanglement (see TELEPORTATION), in which somehow the information between the two particles is transferred. Parapsychologists say that our minds are physical objects and therefore can be described by quantum theory. In other words, like two quantum particles, information can be transferred between minds. So, telepathy can be explained as a quantum connection. Sounds convincing, but there is a major flaw in this argument: our brains are too chaotic to sustain a fragile state known as quantum coherence, which denies quantum entanglement.

Is ESP fact or fantasy? Science can neither prove nor disprove ESP. Some hold the view that if ESP could be supported by empirical evidence, it would no longer be paranormal phenomena. Is there a psychic out there who has won lottery after lottery? No one has ever succeeded in peeking a little way into the future. Until it can be demonstrated convincingly by objective researchers, ESP remains beyond belief.

Two Harvard University objective researchers, Samuel T. Moulton and Stephen M. Kosslyn, recently set out to resolve the paranormal phenomena debate. If ESP or such things exist, they occur in the brain. Moulton and Kosslyn scanned the brains of nineteen pairs of individuals (couples, romantic friends, twins) to assess whether individuals can have knowledge that does not come from normal perceptions. The participants viewed 240 pairs of photographs while inside an fMRI scanner. Each picture pair was randomly assigned a stimulus category – ESP or non-ESP. ESP stimuli pictures were also presented to the subject via three different forms of ESP: telepathically (shown simultaneously to the subject's partner in a separate location); clairvoyance (displayed on a computer located outside the subject's field of vision; and precognition (shown to the subject at a later time).

The researchers found that the participants responded identically to both types of stimuli. The result supports the null hypothesis (the opposite of the hypothesis being tested). 'We didn't find anything, but we didn't find anything in an interesting way,' the researchers say.' They agree that while null results can never be used to conclusively disprove that ESP doesn't exist, but they are happy to have done their bit in settling an age-old debate that belongs on the fringes of science.

Face on Mars

33

Unmasked!

THE IDEA OF intelligent life on Mars was, in fact, born in 1877 when Earth and Mars were in 'favourable opposition' – that is, their orbits had brought them to their closest – something which takes place every 30 years or so. Italian astronomer Giovanni Schiaparelli then noticed that the Martian surface was crisscrossed with a network of about 100 lines. He saw them again in 1879 and again in 1881. He called these lines *canali* ('channels'), which suggested that they were a natural feature. However, the Italian word was mistranslated into English as 'canals'. The idea of such large-scale artificial structures led to wild speculations about intelligent life on Mars (*story 50*).

A century later these speculations became much wilder when in 1976 NASA released a photograph taken by its unmanned Viking spacecraft which was circling the red planet at a distance of 1,860 kilometres (1,162 miles), snapping photos of possible landing sites. The grainy image looked like a shadowy likeness of a human face. To engage the public and attract attention to its Mars missions, the caption was a bit of poetic license: 'The huge rock formation ... which resembles a human head ... formed by shadows giving the illusion of eyes, nose and mouth.'

The Face on Mars soon became a pop icon and an urban myth that it was evidence of past intelligent life on Mars. In 1987 Richard C. Hoagland, an American science journalist, claimed in his book, *The Monuments of Mars: A City on the Edge of Forever*, that the face was the sculpture of a human being located near a ruined city whose 'forts' and 'pyramids' could still be seen. Like Stonehenge, the face pointed to the place where the Sun rose on the Martian solstice about 500,000 years ago when the gigantic face was constructed. Hoagland's book is still in print and propagating the myth.

NASA tells a different story. In 2001 the Mars Global Surveyor spacecraft captured the first high-resolution photo of the face. 'Each pixel in the 2001 image spans 1.56 meters, compared to 43 meters per pixel in the best 1976

Viking photo,' NASA claims. What the picture now shows is the Martian equivalent of landforms common around the American West. There are no eyes, no nose, and no mouth on the 3-kilometre (nearly 2 miles) long and 240 metres wide (about 800 feet) high Face. It's the work of erosion on Martian rocks, not extraterrestrial beings.

But urban myths don't die easily. A statement on the NASA website admits, 'Some people think the Face is *bona fide* evidence of life on Mars – evidence that NASA would rather hide, say conspiracy theorists. Meanwhile, defenders of the NASA budget wish there was an ancient civilization on Mars.' Everyone loves ET.

THE FIRST OF the following two pictures show the grainy photograph taken by the *Viking* spacecraft in 1976 which started the Mars face controversy. The second picture is a high-resolution image of the so-called face on Mars taken by the *Mars Global Surveyor* spacecraft in 2001. (Photos courtesy of NASA)

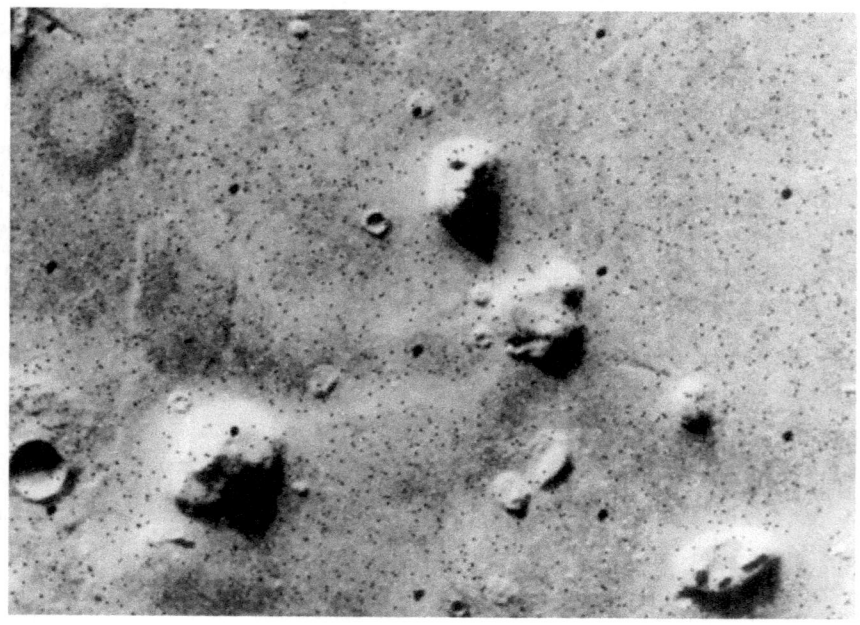

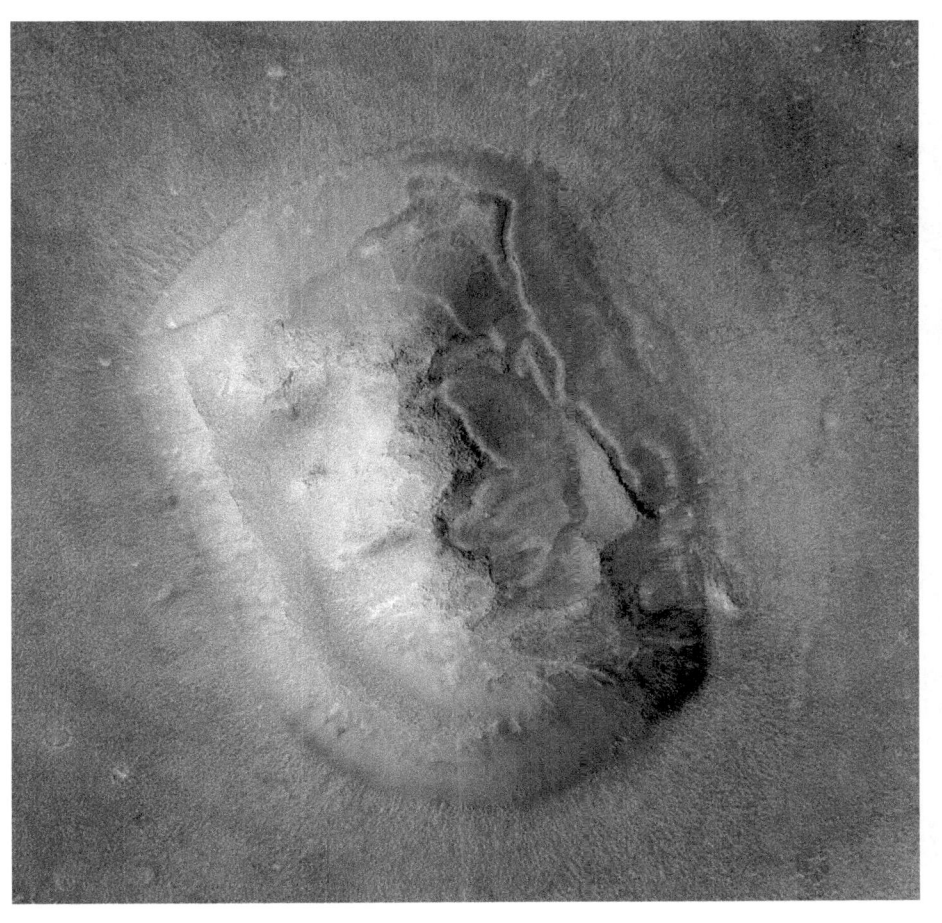

Flat Earth

34
No one has ever fallen off the edge

WHEN CHRISTOPHER COLUMBUS discovered America he proved that Earth is round; before his adventurous voyage of 1492 people believed that Earth was flat and that he might sail off the edge. After more than five centuries this historical fallacy is still firmly established in the public consciousness.

The sixth-century BC Greek mathematician Pythagoras (famous for his theorem) was the first to declare that Earth is a globe. In the second century AD Claudius Ptolemy, the last of the great Greek scientists, gave many reasons for supporting this view, but the most ingenious was that if Earth were flat sunset and sunrise must take place at the same times, no matter what country you were in. He proved that the times of sunset and sunrise changed greatly as the longitude changed. Ptolemy also drew projection maps which replaced plane maps used by ancient geographers.

When Columbus saw the map of the known world drawn by Ptolemy, he was inspired to travel to the East by sailing west. He knew that Earth was round; he hoped that its circumference was small enough to allow him to reach India before his ships ran out of food and drinking water. Columbus' estimate of the circumference was based on Ptolemy's erroneous estimate. Instead, Columbus reached America, calling its native population 'Indians'.

If there was still a shadow of doubt in anyone's mind about the roundness of the Earth after Columbus' voyage, it should have been dimmed when Ferdinand Magellan sailed around the globe in 1519, and completely dissolved by the first photos of Earth from space published in 1969. But even in this century, there are many people who genuinely believe that Earth is flat – it's a disc about 9,000 kilometres (5,600 miles) deep and stretches horizontally forever. The flat Earth theory lives on and the Flat Earth Society is trying its best to nourish it, at least on the Internet.

In everyday language, the term 'flat-earther' is used to describe someone whose beliefs are spectacularly backward and irrational. This has not stopped purveyors of flat Earth myth to claim that Apollo moon landings were faked, as were the photos of Earth from space. It's a big conspiracy to keep all of us in the dark shadow of the flat Earth, they claim, and the simple truth is that Earth is flat because it appears flat. Only if you stick your head in sand!

Four elements

35

Aristotle's gift

THE SIXTH-CENTURY BC Greek philosopher Thales was considered by his contemporaries one of Greece's Seven Wise Men. He was also a keen astronomer and predicted an eclipse of the Sun in 585 BC. One night he was gazing at the sky as he walked and fell into a ditch. A clever and pretty girl lifted him out and remarked sarcastically, 'Here's a man who wants to study stars, but cannot see what lies at his feet.' This incident gave birth to the image of an archetypical absent-minded philosopher (or professor).

Absent minded he might have been, but Thales' scientific ideas were based on observed facts, not myths, though he lived in a world where mythology provided answers to almost everything. He was the first to ask, 'What are things made of?' He answered it by saying that everything in the world – land, air and all living things – had begun as water and would eventually become water once again. What he had proposed was a theory about the origin of things, which competed with the creation myths that were popular during his days.

A century and a half later Empedocles, another Greek philosopher, suggested that everything in the cosmos was made from four elements – earth, water, air and fire. His contemporary, Hippocrates, now known as 'the father of medicine', based his medicine on the balance of these four elements – water (cold and moist), air (moist and hot), fire (hot and dry) and earth (cold and dry) – and four humours (bodily fluids) – phlegm (water), blood (air), yellow bile (fire) and black bile (earth).

In the fourth century BC, Aristotle accepted the notion of four elements and advanced it. He said that each of these elements had its own type of movement, in a certain direction: cold and dry earth went down, hot and dry fire went up, cold and wet water went above earth, and hot and wet air went above water but below fire.

Aristotle is still admired as a great philosopher, but in matters of science he was wrong a lot of the time. Yet for more than 2,000 years his ideas ruled scientific thinking and were accepted without question. There has never been a scientist whose teachings received a kind of divine reverence for so long. His view of elements dominated scientific thinking until it was displaced in AD 1661 by Robert Boyle, the English chemist who established chemistry as a science. In his most famous work, *The Sceptical Chymist*, he rejected Aristotle's notion of four elements and said that elements were 'certain primitive and simple, or perfectly unmingled bodies'. In other words, elements are one of the simplest components of matter, which could not be converted into anything simpler.

In 1869 the Russian scientist Dmitri Mendeleev arranged the sixty-six elements known then in order of their atomic weight (now known as relative atomic mass) and called this arrangement the periodic table. The modern periodic table has 118 elements at present. They are not arranged by atomic weight, but by a much more fundamental quantity, atomic number (the atomic number of an element is the number of protons in the nucleus of one of its atoms).

Aristotle's legacy survives in the term 'the elements', which refers to the weather, especially wind and rain.

Geocentricity

36
Neither science nor religion

FIRST, A BRIEF history of geocentrism (the fiction that Earth is the centre of the universe as opposed to the fact of heliocentrism that Earth and planets revolve around the stationary Sun).

Claudius Ptolemy (*c.* AD 90–170), the last of the great Greek scientists, synthesised the work of earlier Greek astronomers and enhanced this synthesis with the results of his own observations and ideas. But the astute mathematician and astronomer could not detach himself from the great Aristotle's (wrong) idea that heavy objects fall towards the centre of the cosmos faster than the light ones. If Earth were not at the centre of the universe, argued Ptolemy, it would fall towards the centre. Not only this, the massive Earth would fall faster than things on it, leaving everything floating in space. Therefore, he placed Earth at the centre of the universe.

After Ptolemy, Europe had started sinking into the Dark Ages, and the fast-evolving Catholic Church wholeheartedly accepted this view. For fourteen centuries no one dared to challenge the Church's doctrine: we humans and the planet we inhabit are at the centre of the God's creation. With great zeal, the Church would use this doctrine to silence science, again and again.

In the 16th century, the Polish astronomer Copernicus rejected this ancient and commonly held wisdom. To the contrary, he declared that the moon spun around Earth and the moon and Earth together spun around the Sun.

Copernicus completed his book, *De Revolutionibus Orbium Coelestium* ('On the Revolutions of the Celestial Spheres'), which founded modern astronomy, in about 1530, but decided not to publish it. He knew that the book would be seized and destroyed by the Church as it had emphatically displaced man from the centre of God's creation. However, his ideas had become known. Uneducated and pig-headed people merely poked fun at them; learned people, especially religious leaders, were full of venom. Martin Luther denounced him as 'an upstart astrologer, the fool who wanted to turn the whole science of astronomy upside down'. John Calvin quoted, as a higher astronomical authority, Psalm 93 against him, 'God has established the world; it shall never be moved', and asked, 'Who will venture to place the authority of Copernicus above that of Holy Spirit?'

Copernicus passed his life in agony. His efforts to convince the Church of the truth were in vain. At the very end of his life, his pupil and friend Georg Rheticus took the manuscript to Nuremberg and printed about six hundred copies. The first copy of the printed book was sent to Copernicus in Poland. It arrived but a few hours before his death by cerebral paralysis on 24 May 1543. The copy was placed in his hands. The greatest astronomer of his time died without knowing that his book, a monument of scientific genius, had been published.

The book was in defiance of the Bible, which said 'Sun, stand thou still' (Joshua 10:12), but it was too late for the Church to do anything. However, the Church duly declared it 'a false Pythagorean doctrine contrary to the Holy Scriptures,' banned it and placed it on its Index of Prohibited Books. But it was too late; the book had dealt a final blow to the Earth-centred universe of Ptolemy.

A few decades later the Dominican monk Giordano Bruno was burned at the stake and the great Galileo was caged in his house. Their crime? Pushing Copernicus' heretical views zealously. But astronomy would never look back.

It's amazing that there are still people who reject well-established scientific theories and believe that Earth does not move at all. The modern geocentrism began in 1967 when Walter van der Kamp, a Dutch-Canadian school teacher, circulated a paper to some Christian individuals and organisations arguing the case for the doctrine that Earth is the centre of the creation. In 1971 Kamp founded the Tychonian Society, which was named after the famous 16th-century Danish astronomer Tycho Brahe who dissented from the Copernican theory and accepted without question the dogma that Earth stood still at the centre of the universe and the Sun went around it ('The Sun', he declared, 'is the Leader and the King who regulates the whole harmony of planetary dance'). *The Bulletin of the Tychonian Society* dedicated itself to the defence of a geocentric universe and featured articles on the history, philosophy and scientific arguments for geocentrism.

In 1991 the Tychonian Society was reorganised as the Association for

Biblical Astronomy and its bulletin became *The Biblical Astronomer*. The Official Geocentricity Website claims: 'Of all the sciences, the Holy Bible has more to say about astronomy than any other ... This site is devoted to the historical relationship between the Bible and astronomy. It assumes that whenever the two are at variance, it is always astronomy ... that is wrong. History bears consistent witness to the truth of that stance.'

The basis of modern geocentrists' belief is the literal interpretation of the Bible. They may consider themselves the standard bearers for an ill-conceived fusion of science and religion; but science doesn't seek evidence for its theories in religion, and religion doesn't need science's approval for its beliefs.

Graphology

37

Handwriting as character

DOES YOUR SCRIBBLED signature or the way you dot i's and cross t's reveal something about your personality? Proponents of graphology reply with an emphatic 'yes'; but there is no empirical evidence to justify their claim.

Graphology (handwriting analysis) is examination of a person's handwriting to determine personality characteristics. It should not be confused with document examination in which forensic experts inspect documents to detect their authenticity.

Graphology, which has had a long history, is based on the premise that a person's handwriting is influenced by his or her subconscious. By interpreting it properly we can learn about the person's character, traits, attitudes, values and mental abilities. The things graphologists look in a person's handwriting include the pressure exerted by pen on the paper, page layout, spacing of words and letters, loops, ending strokes, dotted i's and crossed t's, slant and speed writing and consistency of style.

Almost all large-scale studies have shown that there is no relationship between handwriting and personality and graphology cannot be considered a useful handmaiden of psychology. However, researchers at Haifa University have claimed that writing with large strokes and applying high pressure on paper could be added to telltale signs that someone might be lying. By analysing these physical properties of students' handwriting, the researchers claim that they could tell whether or not the students were writing the truth. 'Lying requires more cognitive resources than being truthful, you need to invent a story, make sure not to contradict yourself,' the researchers claim. 'Any task done simultaneously, therefore, becomes less automatic. Tabletop pressure sensors showed this effect in the students' handwriting became more belaboured when they fibbed.' This study doesn't exonerate graphology from the accusation of lying; it still remains as phoney as astrology (*story 9*), palmistry (*story 63*) and phrenology (*story 68*).

Great Wall of China

38

The Moon myth

COULD YOU SEE the Great Wall of China from the Moon?

In 1938, well before the advent of space flights, American adventurer and intrepid traveller Richard Halliburton claimed in his book *Second Book of Marvels: The Orient* that the Great Wall is the only man-made object visible to the unaided human eye from the Moon. This started the popular urban myth.

The only thing you can see from the Moon, says NASA astronaut Alan Bean, is a beautiful sphere, mostly white (clouds), some blue (ocean), patches of yellow (deserts), and every once in a while, some green vegetation. From the window of a spacecraft in a low orbit of Earth you can see the Great Wall, the Great Pyramid of Giza and many other man-made objects, if you know where to look and the weather is right. No artificial object is visible from the window of the International Space Station which orbits at about 360 kilometres (224 miles) up; however, you may spot big cities such as London, New York and Beijing in the daytime if you were good in your geography class.

The first-ever picture of Earth from Mars shows our planet blue and beautiful against the deep darkness of space. The picture was taken in 2003 by the Mars Global Surveyor spacecraft from a distance of 139 million kilometres (86 million miles). The most famous picture of Earth was taken in 1990 by the Voyager spacecraft from the edge of the solar system at a distance of 6.4 billion kilometres (3.9 billion miles). In his book, *Pale Blue Dot: A Vision of the Human Future in Space* (1994), Carl Sagan used the picture as a metaphor for the insignificance of our world in comparison to the cosmos. From a distance of 100 or more light years, the aliens would see the 'pale blue dot' in their sky only if they had unimaginably powerful optical telescopes.

Hollow Earth

39
Aliens in the deep

'EARTH HAS A hollow interior' was first proposed as a scientific theory by the English astronomer Edmund Halley (of comet fame) in 1691 at a meeting of the prestigious Royal Society. Halley, who had helped in the publication of Newton's *Principia* in 1687, argued that Earth was composed of four shells: an outer shell 800 kilometres (500 miles) thick, then two inner shells of diameters comparable to Mars and Venus and the solid inner shell about the size of Mercury. He also suggested that the atmosphere between the two inner shells was illuminated and therefore the shells were capable of bearing life. When a brilliant *Aurora borealis* (beautiful display of coloured lights in far northern skies) was seen in 1716, he thought it was caused by the escape of glowing gas in the inner shells through a hole in the North Pole. Obviously, *Aurora australis* (display in far southern skies) was not known to him as Australia had not yet been discovered.

In the 18th century, the famous Swiss mathematician Leonhard Euler expanded Halley's idea, suggesting that Earth was completely hollow, could be reached through holes in the North and South Poles, and had its own sun 960 kilometres (600 miles) wide.

The idea became highly popular when in 1913 Marshall B. Gardner, an American engineer, published a little book, *Journey to the Earth's Interior*, in which he claimed that the extinct mammoth discovered in Siberia in 1846 came from the interior of Earth. Eskimos also originated from the deep, as indicated by their legends of a warm land with perpetual daylight. In 1914 the United States Patent Office accepted Gardner's application for a patent on his hollow-earth theory.

In the 1930s the Nazis, who were suckers for silly ideas, also picked up the idea of hollow Earth. There is no proof but it has been alleged that expeditions were sent to Antarctica and Tibet to meet the aliens in the deep. And there is a bizarre suggestion that Hitler and some other Nazis escaped into the hollow

Earth. True or not, their remains are now definitely there.

The geological truth, however, is not as exciting. Geologists tell us that Earth is made up of three main layers: (1) the outer layer, the crust; (2) below the crust is mantle, a thick shell of molten rock separating the crust from the inner core; and (3) the core is very hot and dense and is about 2,900 kilometres (1,800 miles) below the surface.

You will not find any Nazis or other alien creatures in the centre of Earth, unless you're reading Jules Verne's classic novel, *Journey to the Centre of the Earth* (1864).

40
A medical breakthrough or a major blunder?

WHILE TRANSLATING A medical textbook into German, Samuel Hahnemann, a qualified physician, was puzzled by a passage that described the treatment of malaria by quinine. At that time, no one knew how quinine worked. He decided to experiment with quinine by taking it himself. The drug produced symptoms very similar to those of malaria. Fascinated by this discovery, he started testing other drugs to determine the types of symptoms they produced. To make them safer, he diluted them in alcohol. He was amazed to find that the more dilute a solution was, the stronger its effects.

From his limited observations, which themselves were based on the folk medicine of the day, he came up with the theory 'like cures like'. To describe his method of healing he coined in German the word homeopathy (from the Greek words *homoios*, 'like', and *patheia*, 'suffering'). In 1810 he detailed his new system of medicine in his book, *Organon of Rational Art of Healing*, which is still used today as a basic text of homeopathy.

During those days physicians practised extremely invasive medicine. Nearly every patient had to endure bloodletting by leeches or emetics or purgatives. Hahnemann's little white sugar pills were a boon to them.

There's no scientific evidence supporting homeopathy's principle that a disease can be treated by giving patients ultra-dilute doses of a medicine that would in healthy persons produce symptoms similar to those of the disease. Most large clinical trials have conclusively shown that homeopathic remedies are no better than placebos (*story 70*). If they seem to work, it's because the mind can affect the body's biochemistry. The placebo fools the mind into thinking that the problem is being taken care of.

Homeopathy believers disagree: they say that homeopathy might not work in any known physical way, but it uses forces about which we know very little as, no matter what is the level of dilution, the drug leaves some kind of effect on the water molecules. They are most likely to cite the so-called Belfast homeopathy results which became widely known when the prestigious *New Scientist* magazine included them in its 2005 list of '13 things that do not make sense'. In 2004 Madeleine Ennis, a pharmacologist at Queen's University in Belfast, and her colleagues conducted a study the effects of ultra-dilute solutions of histamine on the basophils, a kind of white blood cells. Basophils produce histamine when we're having an allergic reaction, but once released the histamine stops them releasing any more. The Belfast study showed the amazing result that high dilutions of histamine, probably so dilute that they didn't contain a single molecule of histamine, were able to stop production of histamine by basophils. She claimed that if results turned out to be true, the implications are profound: we may have to rewrite physics and chemistry.

No rewriting of physics and chemistry textbook is necessary as all efforts to independently replicate the Belfast results have failed. In a 2010 edition of the journal, *Homeopathy*, Ennis requests a multi-disciplinary trial to hopefully put an end to this 'never-ending story'. Most scientists believe that the story has already ended.

You can answer the question yourself: Is homeopathy, one of the most popular complementary therapies, a medical breakthrough or a major blunder in the history of medicine?

Perhaps cynical Rick Spleen (British comedian Jack Dee's character in 2007 BBC TV series 'Lead Balloon') has the right answer when he describes homeopathy as 'imaginary cures for imaginary ailments'.

Intelligent aliens

41

Why aren't they here?

THE RECENT DISCOVERIES of hundreds of planets beyond the solar system has made the ages-old question 'Are we alone?' highly tantalising.

Enrico Fermi, the greatest Italian scientist of modern times, was forced to flee Italy in 1938. He moved to the United States where in 1942 he built the world's first nuclear reactor. At a lunch in the summer of 1950 Fermi and fellow nuclear physicists were talking about space travel. The discussion veered towards the possibility of many civilisations beyond Earth. Fermi surprised everyone by asking the provocative question, 'If they are there, why aren't they here?' This is Fermi's paradox.

There are many explanations for Fermi's paradox: *serious* (evolutionary biologist Ernst Mayr: the evolution path that leads to intelligent life is far more complex than we suppose – we are, if not the first, then among the first intelligent life forms to evolve in the galaxy; astronomer Carl Sagan: daunting distances of interstellar space make space travel impossible ... if we are alone in the universe, it sure seems like an awful waste of space); *bizarre* (astronomer John A. Ball's zoo hypothesis (*story 101*) which portrays Earth as a zoo of an intelligent life in the galaxy – they're watching us from a distance); *humorous* (science writer Arthur C. Clarke: 'I'm sure the universe is full of intelligent life – it's just been too intelligent to come here.'); and *optimistic* (astronomer Frank Drake: 'They could show up here tomorrow.').

The English scientist William Whewell, an early voice in the extraterrestrial life debate, said in 1853: 'The discussions in which we are engaged belong to the very boundary regions of science, to the frontier where knowledge ends and ignorance begins.' His words still echo reality after more than a century and a half, when we talk about technology-savvy aliens that have visited us in the past (*story 5*) or have flashed their business cards by radio waves (*story 99*). At least, the search for intelligent aliens is based on this premise, if not the search for extraterrestrial microbial life.

Intelligent design

42

Science or nonsense?

IN 1925 JOHN Scopes, a 24-year-old high-school biology teacher, assigned his students a simple task: reading five pages of a popular biology textbook dealing with evolution. He was charged with violating a State of Tennessee law by teaching the theory of evolution. The law, enacted a few months before, made it illegal to teach any theory that denies 'the Divine Creation of man as taught in Bible'.

The evolution–creationism trial turned the dusty little mining town of Dayton into a carnival. Banners decorated the street. Stalls sold 'Monkey Fizz' lemonade and 'Your Old Man's a Monkey' buttons. Chimpanzees were brought to town to 'testify' for the prosecution. About 1,000 people jammed the courthouse. After seven days of testimony and arguments between the prosecution ('If evolution wins, Christianity goes') and the defence ('Scopes isn't on trial; civilisation is on trial') and eight minutes of deliberations by the judge, Scopes was found guilty and fined $100. A year later, the Tennessee Supreme Court overturned the verdict and dismissed the charges, concluding, 'Nothing is to be gained by prolonging the life of this bizarre case.'

After 80 years, another 'bizarre case' came before a US court. In 2004 the Dover Area School District in Pennsylvania directed teachers to preface the teaching of evolution with a disclaimer saying that evolution isn't a fact. After a 40-day evolution–intelligent design trial, the court ruled in 2005 that intelligent design isn't a scientific theory but a repackaged creationism and banned the District 'from requiring teachers to denigrate or disparage the scientific theory of evolution'.

Charles Darwin first presented his theory of evolution in 1859 in his book, *On the Origin of Species*, in which he said that all present-day species have evolved from simpler forms of life through a process of natural selection. Organisms have changed over time, and the ones living today are different from those that lived in the past. Many organisms that once lived are now extinct. In a subsequent book, *The Descent of Man*, published in 1871, he discussed the idea that humans evolved from apes.

Darwin's ideas were strongly opposed by many contemporary scholars because they conflicted with the religious belief that every species was created by God in the form in which it exists today and is not capable of undergoing any change. However, the movement now known as creation science arose in the last decades of the 20th century. Henry M. Morris, a Baptist trained as a hydraulics engineer, is widely considered its founder. In his books, *That You Might Believe* (1946) and *The Bible and Modern Science* (1951), he proclaimed that the universe was created in six days and Earth's geological features and life forms could be explained by Noah's Flood. Earth was only several thousand years old (not about 4.6 billion years as claimed by scientists) and evolution was therefore impossible. These and other creationism views were based on literal interpretations of popular biblical stories from the Book of Genesis.

In 1989 Foundation for Thought and Ethics, a Christian think tank based in Texas, published a high school textbook, *Of Pandas and People*, which introduced a new label for creationism—intelligent design. Though proponents of intelligent design reject intelligent design as a rebranding of creationism, it remains riddled with the same scientific inconsistencies as creationism.

In brief, creationism/intelligent design holds that God created all things in their present form; they do not share common ancestors. The world we see today is essentially as it was created. Furthermore, as complex intricate features of organisms – at the anatomical, cellular and molecular level – cannot be explained by evolution, we can draw the conclusion that they are the product of intelligent design, not evolution. In support of this argument, the proponents of intelligent design quote Darwin who cited the example of the eye as a complex structure that could not have evolved. Recent researches, however, show that in various families of organisms, eyes have evolved independently.

The proponents of intelligent design often say that evolution is just a theory. As a theory it does such a good job explaining so much of biology that in 1973 Theodosius Dobzhansky, famous Ukrainian-American evolutionary geneticist, was prompted to say that 'nothing in biology makes sense, except in the light of evolution.' A scientific theory tries to comprehensively explain some aspects of nature based on a vast body of evidence; it also predicts testable explanations of phenomena that have not yet been observed. The theory of evolution is based on massive fossil records and countless observations of structures and behaviours of organisms which clearly point out evolution can result in both small and larger changes in populations of organisms. An example of small-scale evolution is that many disease-causing bacteria have been evolving increased resistance to antibiotics. Over long periods of time large-scale evolution can produce organisms that are very different from their ancestors.

Advances in modern biology, especially in knowledge of DNA, have not diminished but enriched the theory of evolution. The same cannot be said for intelligent design.

There is no conflict between religion and evolution. Says Kenneth Miller, an American biologist and the author of *Finding Darwin's God: A Scientist's Search for Common Ground Between God and Religion* (2000), 'Creationists inevitably look for God in what science has not yet explained or in what they claim science cannot explain. Most scientists who are religious look for God in what science does understand and has explained.'

43
The power of distant prayer

INTERCESSORY PRAYER OR distant prayer (prayer on behalf of others) is widely believed to help recover patients from illness. Sadly, no well-controlled clinical trial so far has supported this belief.

In a scientifically rigorous study, which lasted almost a decade, a team of sixteen American researchers monitored 1802 patients at six hospitals nationwide for 30 days after the patients had undergone heart surgery. They evaluated whether (1) receiving intercessory prayer or (2) being certain of receiving intercessory prayer was associated with recovery without any complications after the surgery. The patients were randomly assigned to one of the three groups:

- Group 1 (prayer but uncertain): 604 patients received intercessory prayer after being told that they may or may not receive prayer.
- Group 2 (no prayer but uncertain): 597 patients did not receive intercessory prayer also after being told that they may or may not receive prayer.
- Group 3 (prayer and certain): 601 patients received intercessory prayer after being told they would receive prayer.

The prayers were delivered by members of three congregations using the patients' first name and the first initial of their last name. The prayers were provided for 14 days, starting the night before the operation. After analysing complications in the 30 days after the heart operation, the research team found no differences between those patients who were prayed for and those who were not.

In fact, patients who knew they were being prayed for had more complications those who did not know. Fifty-nine percent of Group 3 patients who knew they were being prayed for suffered complications compared with 52 per cent of Group 1 patients who were not. The researchers believe that strangers' prayers may have made patients worried that their conditions were worse than they have been told: 'It may have made them uncertain, wondering am I so sick they had to call in their prayer team.'

Major complications and deaths were similar across the three groups.

Though the study report was released in 2006, it still remains the most comprehensive study of distant prayer.

No study can answer the question of whether we should pray for the recovery of our sick loved ones. Unfortunately, medical science does not follow the precise laws of the universe as described in physics textbooks. We must pray (without telling the patient), but should not expect that by some unknown law of quantum mechanics our prayers will generate cosmic waves that will improve the physical wellbeing of the patient. At least, prayer will make you feel happy.

Killer asteroids

44

Should we lose sleep over asteroid threat?

ASTEROIDS ARE NOW star attractions for astronomers around the world. This attention is worthy of their name, which is Greek for 'starlike', but it has more to do with their sheer number and their destructive power than their cosmic beauty. These pockmarked, giant, peanut-like rocks are in fact leftovers from the formation of the planets, and they orbit within a vast, doughnut-shaped ring between Mars and Jupiter, known as the main belt.

Occasionally, a collision may kick an asteroid out of the belt, sending it onto a dangerous path that crosses Earth's orbit. These stray asteroids take up an orbit that loops past Earth, and are called 'Earth-crossers'. This knowledge frightens astronomers. In recent years astronomers have identified hundreds of Earth-crossers one kilometre or more in diameter and hundreds more await discovery – there may be as many as 100 million Earth-crossers larger than 20 metres (65 feet).

What if one of them comes too close to Earth? What cataclysm would such a rogue rock cause if it slams into Earth? The number of asteroids is very large, but the space they occupy is enormous. Most asteroids stay millions of kilometres apart. It's not like *Star Wars* or *Star Trek* spaceships weaving their way through flying rocks. But real collisions are possible with a spacecraft or Spaceship Earth.

The Tunguksa explosion (*story 95*) was probably caused by an asteroid. Are we going to be hit by a Tunguska-like asteroid again? Astronomers suggest that the average frequency of impacts of this size (average width 75 metres or 246 feet) is 1 in 1,000 years. These asteroids explode in the lower atmosphere but release enough destructive energy to wipe out a large city. Asteroids less than 10 metres (32 feet) wide may not survive Earth's atmosphere.

The rate of impact decreases with the increase in size of the asteroid. The average interval between impacts of giant asteroids (average width 16 kilometres or 10 miles) is 100 million years. An impact of this magnitude could cause a global effect, event extinction, in part by spreading dust cloud that could blanket the most of the planet. The impact of an 8-kilometre (5 mile) wide asteroid on the Yucatan Peninsula of Mexico is believed to have wiped out the dinosaurs around 65 million years ago.

Do we really want to win an asteroid lottery? The probability of randomly picking six numbers in a lottery of 45 numbers is 1 in 4 million. The probability increases to 1 in 14 million if you have to randomly pick six numbers from 49 numbers, and to 1 in 19 million if you have to pick from 51 numbers. The probability of an asteroid impact, small or big, is 1 in 20,000, the same probability as for a passenger aircraft crash. From these odds it appears that the proverbial man in the street, if he is not run over by the proverbial bus (probability 1 in 100), will witness an asteroid impact long before he wins the big lotto. Why bother to buy a lottery ticket today?

Should we dismiss these risk probabilities as lies, damned lies and statistics, or lose sleep over the asteroid threat? Experts say that the risk isn't large enough to keep us awake at night, but it isn't completely negligible either. What should an ordinary person do? NASA advises as follows: The next time you see a headline '*Killer asteroid threatens Earth?*' ask yourself two questions: (1) Have you known about this space rock for more than a week or so? If not, check again in a month. It won't be considered a killer then. (2) What are the odds of impact? If you're likely to win the lottery, there's probably nothing to worry about.

And a killer asteroid didn't end the world on 21 December 2012 (*story 51*), neither will planet Nibiru (*story 71*) or Pole reversal and pole shift (*story 73*).

Kirlian photography

45
Images of fringes, not halos

IN 1939 SEMYON Davidovich Kirlian, a Russian electrician, discovered that human bodies – all living things, for that matter – radiate energy when stimulated by high-frequency, high-voltage alternating current. He called this bioplasmic energy, a kind of energy field surrounding all living things. Kirlian claimed that bioplasmic energy came from oxygen. The deep breathing recharges the body and helps to distribute energy to all parts of the body; and breathing negatively charged ionised air was highly effective in relieving mental and physical tension.

The Kirlian photography setup is simple. A photographic film is placed on the top of a metal plate, and the object to be photographed (usually fingers, in the case of humans) is placed on the top of a film. A high-voltage pulse is passed momentarily through the metal plate to make an exposure. When the film is developed you have a Kirlian photograph, which is indeed striking, showing a whole new world of colours and patterns. The most famous effect of Kirlian photography occurs when a freshly cut plant leaf is photographed, then torn in half and photographed again. A faint image of the torn-out piece can still be seen in the second photo.

Paranormalists embraced this phenomenon whole-heartedly and claimed that Kirlian photographs were physical proof of the existence of human spiritual aura. This view of Kirlian photography was introduced to the Western world in the early 1970s when Sheila Ostrander and Lynn Schroeder published their book, *Psychic Discoveries Behind the Iron Curtain*. They made ostentatious paranormal claims about Kirlian photographs: in humans the energy field varied according to the person's health and state of mind; and the photographs could be used to foretell illness before there were any signs of disease

Scientific investigations of Kirlian photography have revealed that the photographs do show colourful fringes around the borders of living things; however, such fringes are not present in photographs of nonliving things. Fringes in the photographs of living things, scientists explain, are caused by moisture present in the object to be photographed. When an electric discharge enters the object, it ionises the area around the object. During exposure the moisture is transferred from the subject to the film, causing a pattern of electric charge on the film. This pattern is enhanced when photographing, say, fingers of a person who is emotionally disturbed. Emotional disturbance causes sweating which, in turn, enhances the fringe pattern of the photograph. When the photographs are taken in a vacuum, the effect disappears. If Kirlian photographs were surely showing some paranormal phenomenon, we wouldn't expect the phenomenon to disappear in a vacuum (it was Aristotle who said nature abhors a vacuum; he probably didn't know that paranormal entities also abhor a vacuum).

James Randi, famous for debunking many paranormal myths, writes in his *An Encyclopedia of Claims, Fraids, and Hoaxes of the Occult and Supernatural* (1997): 'Once highly regarded by the paranormalists, Kirlian photography has now been shown to only indicate variance in pressure, humidity, grounding, and conductivity. Corona discharges are well understood and explained in elementary physics.' The problem is that those who believe in things paranormal do not read physics textbooks, however elementary.

'Left brain' and 'right brain' myth

46
A tale of two brains

THE FAMOUS OPENING sentence in Charles Dickens' 1859 novel, *A Tale of Two Cities*, 'It was the best of times, it was the worst of times, it was the age of wisdom, it was the age of foolishness …', aptly applies to this modern tale of two brains.

In this age of wisdom, many of us still believe in the foolish idea that rationality, logic and verbal skills are located in the left hemisphere of the brain, while creativity, emotions and visuospatial skills are located the right hemisphere. This assumption leads us to believe that left-brained people are logical and good at mathematics and right-brained people are artistic and bad at mathematics.

The erroneous thinking that information is processed in different ways in the two hemispheres of the brain is still reflected in our schools: the best teaching techniques for left-brained people should involve verbal instructions, talking and writing, and multiple-choice questions; while demonstrated instructions, drawing and manipulating objects, and open-ended questions are best for right-brained people. This notion has led to the idea that education programs should synchronise the two hemispheres by including both left-brained and right-brained activities. 'Show and tell' activities of your early school days are the result of this thinking: instead of only reading a 'left-brained' text, your teacher also showed pictures and graphics to stimulate your right hemisphere.

The left brain/right brain myth can be traced back to the days of the 19th-century craze of phrenology. Phrenologists believed that different mental functions were located in different organs of the brain, and the growth of the various organs was related to the development of associated mental abilities. As this growth would be reflected in the shape of the skull, personality traits could be determined by reading bumps and depressions on the skull.

In 1844 this mumbo jumbo became popular when a book described the two hemispheres of the brain as independent parts having an independent way of thinking. The idea even found a way into Robert Louis Stevenson's famous story, *The Strange Case of Dr Jekyll and Mr Hyde*, published in 1886.

In the 1960s the myth found its way into the modern scientific literature when American scientists Roger Sperry, Joseph Bogen and Michael Gazzaniga embarked on what are now known as split-brain studies: how the brain's left and right hemispheres are specialised for different tasks. Their conclusion was based on the study of patients, usually, epileptic, who had undergone a surgical procedure that severed the whiter matter neural fibres that link the two hemispheres of the brain.

However, in the hands of psychologists these findings took life of their own. In his 1972 best-selling book, *The Psychology of Consciousness*, psychologist Robert argued that we place too much emphasis on rational, left-brain thinking and not enough on intuitive, right-brain thinking. Psychologist Betty Edwards's *Drawing on the Right Side of the Brain* stressed the benefits of creative, right-brain thinking.

The split-brain research has now moved from a static view of what happens in a particular hemisphere to a much more interactive view how the whole brain, interacting through white matter fibre systems, orchestrates the entire cerebral network into coherent and apparently seamless cognitive action.

It amuses Gazzaniga that his split-brain work has now become such a part of popular culture. He laments that it has become 'mixed up' with sound psychological and educational work that demonstrates that children use a variety of cognitive strategies to solve problems.

'There are some kids who visualise problems and other kids who verbalise them,' he says. 'That reality has been mapped on left brain/right brain anatomy as an explanation. But that's where it falls down. Cognition, in general, is much more complex than that. That's what we have learned over the years and continue to learn as we study hemispheric differences.' In brief, the brain's two hemispheres do not work independently; they work in a highly coordinated fashion.

Sophisticated brain-imaging techniques also reveal less romantic sides of the brain: there is no evidence that the left brain is 'mathematical' and the right brain 'musical'. Yes, the brain is divided into two hemispheres. They look almost identical anatomically, but they are not independent. They are connected by thick bundles of nerve cells which carry information from one side to the other.

The two hemispheres differ not so much in what they do, but in how they process tasks. The left hemisphere is better at details (such as recognising a particular face in a crowd), whereas the right hemisphere is better at dealing with a general sense of space (the relative positions of people in a crowd). In the case of language, for example, the left hemisphere focuses at step-by-step processes, such as grammar and word generation, whereas the right hemisphere focuses at feeling a rhythm, such as intonation and emphasis of speech.

There are no specific 'left brain' or 'right brain' cognitive functions. Both hemispheres work in concert with each other, whether we are reading, painting or solving an algebra equation. It's time we used our whole brains to learn that like Chinese Yin Yang symbols the two hemispheres of our brains are in perfect harmony.

Don't let the left brain/right brain myth stereotype children's capabilities and limitations. Individuals do have relative strengths and weaknesses but it doesn't mean that we let Jane think that she is not good at mathematics because she is right-brained. It would be foolish to think that our strengths and weaknesses come from the dominance of one hemisphere.

There is no program or technique that can boost capabilities of your right or left brain. Similarly, no scientific study supports the claims made by 'whole brain' training programs. Why waste money on such programs to exercise your brain, when you can exercise your brain on your own for free by learning a new language

47
It's time you met Monsieur Litre and Mademoiselle Millie

IN THE LATE 18th century, science in France was developing faster than anywhere else in the world, but the French employed more units than any other country. For length, for example, they had the *ligne* (somewhat longer than the old British line, which was 1/12 British inch), twelve *ligne* made a *pounce* or French inch, twelve *pounce* made a *pied* or French foot, six *pied* made a *toise*, and then 3,000 *toise* equalled a French league. To end this confusion, the French Academy appointed in the middle of the French Revolution a commission of twenty-eight eminent scientists. In 1790 the commission standardised the units of length, weight and volume and named them *metre*, *gram* and *litre*. The units became official in 1799.

The metric system we use today is known as the SI (Système International) and was introduced in 1960. Many SI units such as Ampere, Celsius, Joule, Newton, Ohm, Pascal and Volt are named after scientists, but litre is definitely not named after the fictional 'French scientist Claude Émile Jean Baptiste Litre (1716–78)'. In 1978, *Chem 13 News*, a Canadian chemistry newsletter, published a 'biography' of 'Litre' suggesting that 'in celebration of the 200th anniversary of the death of this great scientist, it has been decided to use his name for the SI unit of volume (the abbreviation will be L, following the standard practice of using capital letters for units named after scientists)'. The spoof turned into a literary hoax when a précis of the article appeared in the newsletter of the prestigious International Union of Pure and Applied Chemistry. The spoof also fooled many other publications and radio programs. But most scientists got the joke. Some joined in the fun, creating a daughter for Litre, named Millie.

The metric system is based on the decimal (base-10) system and is now used widely, except in the United States.

Loch Ness monster

48

We see what we want to see

MONSTERS, MYTHICAL OR real, fascinate everyone. But no creature in modern times has fired the popular imagination as the Loch Ness monster, affectionately known as Nessie. Its alleged home is Loch Ness (the word 'loch' is Scottish for lake) in the Scottish Highlands. This 290-metre (950-feet) deep lake is one of Britain's largest lakes and, as the prestigious journal *Nature* once scorned, 'the underworld of fables'.

The fable of Nessie began in 565 when a giant serpent-like creature jumped out of Loch Ness and lunged at one of the monks accompanying St Columba on his mission to convert Scotland to Christianity. The monster disappeared when the good Irish saint made the sign of the cross. Nevertheless, it never disappeared from the public consciousness, as recurring reports that a monster dwells in the dark waters of Loch Ness continued to tantalise.

It leaped into fame in 1933 when the London *Times* published a detailed story on fifty-one eyewitness accounts and drawings of the monster collected by Robert T. Gould, a retired navy officer. He described the monster as a 15-metre (50-feet) long-necked creature with one or two humps and at least two, possibly four, fins or paddles. As the myth of the monster grew so did the hoaxes and fake photographs. All this hoopla led the famous anthropologist and anatomist Arthur Keith (who has now been linked to Piltdown man, *story 69*, to write in 1934 to the *Daily Mail*: 'The existence or nonexistence of the monster is not a problem for zoologists but for psychologists.'

If there was indeed a monster in Loch Ness, how did it get there? Believers say that Nessie-like creatures could have been trapped when the lake was cut off from the sea at the end of the last ice age some 20,000 years ago. Sceptics argue that if there has been a monster in the lake for so long there would have to be at least twenty animals in the herd to support continued breeding over centuries. There is hardly any food in the lake to support that many creatures.

All so-called photos and videos of the monster suggest that it might be a plesiosaur. It could be speculated that these marine reptiles have adapted to the cold waters of Loch Ness despite their preference for sub-tropical waters. But plesiosaurs disappeared with the dinosaurs.

In 2003, a team of researchers assembled by BBC used 600 separate sonar beams and satellite navigation technology to survey the waters of Loch Ness but found no trace of the monster. The BBC team concluded that the only explanation for the persistence of the myth of the monster is that people see what they want to see. The Scottish tourist industry is not complaining, even if Nessie remains elusive forever.

THE FOLLOWING PICTURE: In April 1934 a London gynaecologist Robert K. Wilson allegedly snapped a photo which became one of the most famous photographs of the Loch Ness monster. Wilson later admitted that the photo was a hoax and that he never believed in the creature.

Magnetic therapy

49

Attractive claims, sham benefits

MAGNETS HAVE BEEN touted as a treatment for many medical conditions since the time of ancient Chinese, Egyptian, Greek and Indian physicians. Modern magnetic therapy began when Franz Mesmer mesmerised the 18th-century Paris with his claims of 'animal magnetism' (*story 53*). Though his spurious claims were demolished by a distinguished scientific panel, magnets never lost their power to attract the gullible. Now New Age magnetic therapies claim to cure conditions ranging from back pain to cancer. Their so-called therapeutic devices are also equally wide-ranging: magnetic bracelets, wrist and knee bands, back and neck braces, inner soles of shoes, pillows and mattresses; in fact, any device that helps to place a magnet close to your body.

If you ask medical experts the question whether there is anything in the human body that is affected by a magnet, their unequivocal answer would be an emphatic 'no'. Do claims about the healing powers of magnets have any scientific basis? In 2006 two respected American scientists, Leonard Finegold, a professor of physics, and Bruce L Flamm, a professor of medicine, combined their expertise to answer this question. Their review of the medical literature on magnetic therapy found very little supporting evidence. They concluded in the British medical journal, *BMJ*: 'Patients should be advised that magnetic therapy has no proved benefits. If they insist on using a magnetic device they could be advised to buy the cheapest – this will at least alleviate the pain in their wallet.' They also warned that the self-treatment with magnets risked leaving underlying medical conditions untreated. Magnetic therapy may appear harmless, but it doesn't mean it is safe.

Wearing copper bracelets or magnetic wrist bands for relieving arthritic pain is also very popular. They are generally considered safe but almost all medical studies have found little evidence that they provide any real benefits. Any perceived benefits can be attributed to psychological placebo effect (*story 70*).

Martian life

50
Smart Martians and silly scientists

EVER SINCE GALILEO caught sight of Mars through his newly built telescope in 1609, the red planet has been a source of intrigue and imagination. Until the mid-20th century scientists seriously believed that intelligent life existed on Mars.

It has been said of Carl Friedrich Gauss that almost everything that the mathematics of the 19th century brought forth in the way of original scientific ideas is concerned with his name. The question of life on Mars did not escape his attention either. At that time it was believed that Martians could look at Siberia. To attract their attention, Gauss proposed clearing stretches of forest there to form a gigantic right-angled triangle with squares on each side, planting wheat in the triangle, and leaving squares of trees around it. He believed that his diagrammatic demonstration of Pythagoras' theorem would reveal our presence to our neighbours (how could Martian masterminds miss the mathematical message hidden in Pythagorean crop circles, *story 22*).

A few years later, Johann Joseph von Littrow, an Austrian physicist, worried about the poor visibility of Pythagoras' diagram in the dense Siberian forest, suggested that the right thing to do would be to dig giant ditches in the Sahara in the form of geometric figures such as triangles, squares and circles, fill them with water, and pour kerosene on top of the water and set them ablaze at night. Even this bright idea was not followed up.

The idea of a blazing Sahara as a beacon to attract Martians' attention appeared again in 1880 when French astronomer Camille Flammarion suggested that, if chains of light were placed on the Sahara on a sufficiently generous scale to illustrate Pythagoras' theorem, intelligent Martians might conclude that there was intelligent life on Earth.

These ideas certainly fired up the imagination of many 20th-century scientists. In 1920 Guglielmo Marconi said that he had received mysterious radio signals that he believed might have come from Mars or some other region of the universe where electrons are in vibration. He ignored these signals, saying, 'I'm concerned enough at present with business upon Earth.' The *International Herald Tribune* reported, 'Mr. Thomas A. Edison, commenting on the statement of Marconi that untraced wireless calls might come from Mars, stated that such a thing is possible. "Existing machinery is able to send signals to Mars," said Mr. Edison. "The question is, have the beings there instruments delicate enough to hear us? They say Martians are as far ahead of humans as we are ahead of chimpanzees. If that is true they must have such apparatus."'

In 1924 David Todd, an American astronomer who like Edison believed in the superiority of Martian technology, convinced the US government to turn off high-powered radio transmitters on 23 August, when Mars came closest to Earth. On that day, when the two worlds were just 55 million kilometres (34 million miles) apart, transmitters were turned off for five minutes before each hour, so that Todd could listen to Martian chatter. Martians were also aware of Earth's closest approach. On that day they decided to turn off their radio stations and sit in silence.

Todd did not give up his hope of communicating with Martians that easily. He suggested filling a 15-metre (49-foot) bowl with mercury at the bottom of an abandoned Chilean mine and placing a powerful light source at the focus of this natural parabolic reflector. This time he failed to convince authorities to act on his good advice.

1920 was a vintage year for scientific ideas about communicating with Mars. Pages and pages of *Scientific American* for that year are filled with suggestions not only for technical means of signalling but also about the 'language' of the message. 'Should we signal in English or in French or in what language?' asks a reader. 'Or perhaps the enthusiastic mathematicians who tell us, with delightful generality, that mathematics is the universal truth and that therefore the message will be such as to convey some fundamental mathematical idea, will deign to work out their scheme for us, and tell us with just what mathematical truth we are to startle the Martians?'

Someone suggested giving Martians an IQ test by sending two signals, then four, and if they answered back with eight signals we would know that Martians could multiply.

Not everyone was convinced of the existence of Martians. One reader wrote that if all the letters of the Lord's prayer were thrown into the air, there would be as much chance of the letters 'falling back into their proper places to print the prayer without an error as there is of there being inhabitants on the planet Mars with whom we might by any possibility communicate'.

The last serious proposition to attract the attention of Martians came in 1942 from the famous British astronomer James Jeans. He suggested shining a group of searchlights towards Mars and emitting successive flashes to represent series of numbers. If, for instance, he said, the numbers 3, 5, 7, 13, 17, 19, 23 … (the sequence of prime numbers) were transmitted, the Martians might surely infer the existence of intelligent life on Earth.

Mayan calendar

51

The end of a cycle

THE SAXTON STATE Library in Dresden, Germany, founded in 1556, is a treasure trove of medieval and Renaissance manuscripts. In 1739 the library acquired a rare manuscript, one of the only three Mayan manuscripts in existence. Known as the Dresden Codex, it's now the library's most famous and valuable possession. Made from the bark of the wild fig tree, this 3.56 metre (about 12 feet) long manuscript consists of thirty-nine pages which fold up like an accordion. Between 1200 and 1250, eight scribes using single-haired brushes painstakingly painted both sides of the pages with hieroglyphs, icons and scenes of deities performing rituals. Covered in jaguar skin, the manuscript is believed to be a sacred book.

The manuscript has been deciphered as astronomical tables, writings about constellations and planets and predictions of lunar eclipses. The most exquisite six pages of the manuscript, painted in black, red, blue and yellow, are devoted to intricate charts mapping the positions of Venus, the planet of greatest interest to the Maya. Maya priests used astronomical tables to make prophecies and to study the influence of heavenly bodies on human affairs.

The Mayan civilisation thrived for nearly 2,000 years in the lowlands that today comprise Guatemala, Belize, El Salvador and Mexico's Yucatán peninsula. There they built great cities with palaces, plazas and flat-topped pyramids with temples atop them. Besides magisterial architecture, they also created works of astronomy, mathematics and literature that equalled, if not surpassed, the achievements of other contemporary civilisations. Between 800 and 900 this great civilisation of about 6 million people suddenly collapsed. What has baffled archaeologists most is that the collapse was not caused by a single catastrophic event. Was it because of epidemic, drought, overpopulation, war or political breakdown? No one knows.

From the Dresden Codex we learn that the Mayans used three different forms of calendar to track longer periods of time: a 260-day sacred round, a 365-

day solar year and a 52-year calendar round. The origin of the 260-day calendar can be traced back to 900 BC. This calendar is not based on an astronomical cycle, but it corresponds to the human gestation period of nine lunar months. This calendar paired cycles of twenty named days with thirteen numbers to give 260 unique days. The 365-day calendar ('vague' year, as it approximated the solar year of 365.2422 days) was introduced in 550 BC and was divided into eighteen months of twenty days each plus an added five-day period. As the 260-day and 365-day cycles repeated themselves every 52 vague years exactly, this period was known as the calendar round (the least common multiple of 260 and 365 is 18,980 days).

In the second century AD, the Mayans realised that they needed a better way of recording historical events because specific dates within their 52-year or 18,980-day calendar were becoming confused. This new calendar is now known as the long count. The long count recorded a continuous and chronological record of the passing of time from a fixed moment in time (4 Ahaw 8 Kumk'u for the Mayan, and 11 August 3114 BC for us). Curiously, the creators of the long count seem to have also keyed in an 'end' date. Like an odometer the calendar will revert to zero on 4 Ahau 3 Kankin and a new cycle will begin. This date translates to 21 December 2012 (the winter solstice in the northern hemisphere) in the Gregorian calendar we use.

This wsas the basis for the misguided belief that the world would end on 21 December 2012. The ancient Mayans celebrated the end of a cycle. The New Age doomsayers simply exploited the naive and gullible and laughed all the way to the bank, if the numbers of books, websites, documentaries and at least one major movie, *2012*, were any guide.

52
Evolution by imitation

IN HIS 1976 book, *The Selfish Gene*, the English evolutionary biologist Richard Dawkins (now best known for his controversial book, *The God Delusion*, 2006) noted that genes are replicators and evolution can also be based on other replicators. He suggested that like genes that transmit the traits of organisms, there is another replicator which copies and transmit behaviour and ideas from person to person: 'Cultural transmission is analogous to genetic transmission in that, although basically conservative, it can give rise to a form of evolution.'

He called this unit of cultural transmission meme (from Greek *mimema*, 'that which is imitated'). 'Just as genes propagate themselves in the gene pool by leaping from body to body via sperms or eggs,' he said, 'so memes propagate themselves in the meme pool by leaping from brain to brain via a process which, in the broad sense, can be called imitation.' In other words, a meme (pronounced 'meem') is an elementary knowledge entity (an idea, behaviour, style or usage) that spreads from person to person within a culture by nongenetic means, especially by imitation.

Susan Blackmore, an English psychologist who is infected by the idea of meme (which is itself a meme) and attempts to infect people's minds with meme by her articles and the book, *The Meme Machine* (1999), believes that memes have been (and are) a powerful force shaping our cultural – and biological – evolution. 'We may embellish a story, forget a word of the song, adapt an old technology or concoct a new theory out of old ideas,' she says. 'Of all these variations, some go on to be copied many times, whereas others die out. Memes are thus true replicators, possessing all three properties – replication, variation, selection – needed to spawn a new Darwinian evolutionary process.'

Not everyone gets infected by Blackmore's enthusiasm for this new Darwinian evolutionary process. One of those infected by opposing memes is American evolutionary biologist Massimo Pigliucci. He says that it is pretty much impossible to tell what constitutes a meme. A gene, on the other hand, has a physical basis: roughly speaking, they are pieces of DNA. 'Ideas clearly do evolve, and there is in fact a somewhat undeniable analogy between memes and the evolution of genes,' he argues. 'But we don't need to push that analogy too far, and we certainly don't need a whole new vocabulary to make sense of it.'

English psychobiologist Henry Poltkin brings a psychologist's perspective to memes. He argues that culture is not simply a collection of memes, and that all memes, and thus all aspects of culture, are not spread by imitation. 'Human culture is about the sharing of knowledge, beliefs and ideas,' he asserts. 'Imitation, properly defined, does not come into it.'

The meme debate continues. If any of the above ideas of Dawkins, Blackmore, Pigliucci and Poltkin have leapt to your brain, you've fulfilled a requirement of cultural evolution by memes, imitation. But, some critics of meme argue, imitation is useless without creativity.

Mesmerism

53

'The art of increasing the imagination by degrees'

IN 1775 FRANZ Mesmer, an Austrian physician who got his doctorate for a thesis on how the gravity of various planets affects health, proposed that an invisible magnetic field flows through our bodies. If the flow were restricted somehow, it would cause physical and mental illnesses. He theorised that by passing magnets over the body, the fluid would be unblocked and the patient cured. He managed to cure some patients by this method. He eventually discovered that he could achieve the same results by the passing of hands over the body, even when at a considerable distance. He claimed that he was now making use of 'animal magnetism'.

He wrote an account of his discovery to all the learned societies of Europe requesting them to investigate it, but no one took him seriously. Still, he was not discouraged and continued to talk to anyone who would listen to him about his amazing power to magnetise anything. 'I have rendered paper, bread, wool, silk, stones, leather, glass, wood, men and dogs – in short everything I touched, magnetic to such a degree, that these substances produced the same effect as the loadstone on diseased persons,' he once wrote to a friend. 'I have charged jars with magnetic fluid in the same way as is done with electricity.' (He was referring to Leyden jars which were popular during his days and were used to store static electricity.)

Viennese looked upon his pretensions with contempt or indifference. In 1778 he decided to move to Paris as he thought 'Paris, the idle, the debauched, the pleasure-hunting, the pleasure-loving' was the scene for a philosopher like him. He hired a sumptuous apartment and opened his healing saloon. Animal magnetism, or, as some called it, mesmerism, soon became quite the rage, especially among the well-heeled women.

His assistant 'magnetisers', generally strong, handsome young men, would request the patients sit with their feet in a huge tub filled with 'magnetised water' while holding iron rods attached to the tub with one hand and with other hand holding each other's thumbs between their thumb and forefinger. Mesmer would then emerge from behind heavy drapes, dressed in a long robe of purple silk richly embroidered with gold flowers and bearing in his hand a white magnetic rod. He would ask the patients to squeeze their thumbs and forefingers to allow magnetism to flow freely through the group.

This showmanship infuriated Parisians so much that in 1784 King Louis XVI instituted a scientific enquiry into 'animal magnetism'. The royal commission included the renowned French chemist Antoine Lavoisier and the American statesman-scientist Benjamin Franklin, who was visiting Paris at the time. The commission concluded that the observed effects could be produced without magnetic manipulations, and animal magnetism did not account for the phenomena. Furthermore, any observed effects could be attributed to the power of suggestion (a kind of placebo effect, *story 70*) and that the practice was nothing but 'the art of increasing the imagination by degrees'. In other words, the effect on patients was mental, not magnetic. This report ruined Mesmer's reputation in France. He retired to his own country, where he died in 1815.

The well-known science writer Isaac Asimov says that Mesmer was 'ninety percent gobbledygook'. Anyway, he is immortalised in the phrases 'animal magnetism' and 'mesmerise' (for example, 'his roguish animal magnetism mesmerised her') and his medical quackery survives in the New Age magnetic therapy (*story 49*).

Mirror matter

54
Looking glass worlds

WELCOME TO THE mirror matter – a world of mirror planets, mirror stars, mirror galaxies and even mirror life, all governed by mirror forces. This world is as fanciful as the one Alice entered *Through the Looking Glass*.

Because we are made of ordinary matter, we can neither see or smell mirror matter (or our mirror matter twins, even if they are dressed in their brightest mirror matter clothes and soaked in mirror matter perfume). If you did encounter your mirror matter twin, you would pass right through him or her. You would also be invisible to your twin.

There is no evidence that we are surrounded by a parallel universe of mirror matter and scientists hotly debate even the existence of mirror matter.

But what is this exotic matter? Scientists say that all fundamental particles such as electrons, protons and neutrons have an antimatter twin. The first antiparticle, the positron (or the positive electron; it has the same mass as the electron but opposite charge), was discovered in 1932.

The existence of antimatter has led to the idea that every particle also has a mirror-like twin. This means that in addition to the antimatter world, there might also exist a mirror world. In this world, particles are mirror images of ordinary particles. They also have the same mass as their ordinary counterparts.

Mirror matter is still in the realm of science fiction, but it doesn't stop scientists speculating about its uses. When a particle collides with its antimatter twin they annihilate each other, producing an awesome amount of energy. Unlike antimatter, mirror matter doesn't interact with ordinary matter, but some scientists speculate that if hypothetical mirror matter could be captured, it could then be used to generate near-endless amounts of energy. At the moment this idea has as much scientific credibility as the idea of perpetual motion machines (*story 67*).

55
Americans didn't land on the Moon, bat-men did

YOU HAVE PROBABLY heard about the conspiracy theory that the Apollo 11 moon landing was an elaborate hoax. Conspiracy theorists' websites and blogs revved up in 2001 when Fox TV aired a program 'Conspiracy Theory: Did We Land on the Moon?', arguing that NASA technology in the 1960s wasn't up to the task of real Moon landing; instead it faked the Moon landing in movie studios. A witty comment on the NASA website should silence the peddlers of Moon hoax stories who ignore the overwhelming evidence: 'Fortunately the Soviets didn't think of the gag first. They could have filmed their own fake Moon landing and really embarrassed the free world.'

In the early 19th century, when there were no conspiracy theorists around but hoaxes abounded, there was a clever Moon hoax which claimed that a famous British astronomer had discovered intelligent life on the Moon. Here's the story of that hoax.

On 25 August 1835 a New York newspaper, *The Sun*, ran a front-page story with the headline 'great astronomical discoveries lately made by Sir John Herschel'. The story, purporting to be a reprint from a supplement of the *Edinburgh Journal of Science*, described fantastic sights of the Moon viewed by Herschel from his new telescope at the Cape of Good Hope. The story, which continued in instalments over the course of a few days, described the lunar landscape of vast forests, inland seas and lilac-hued 'very slender pyramids, standing in irregular groups, each composed of about thirty or forty spires'.

As *The Sun*'s sales soared from 8,000 to 19,360 copies its readers were introduced to *Vespertilio-homo* or 'bat-man' on the Moon: 'Certainly they were like human beings ... They averaged four feet in height, with short and glossy copper colored hair and had wings composed of a thin membrane ...'

At the time of the publication of the articles, John Herschel, one of the most famous scientists of his time, was in Cape Town surveying the southern skies (John was son of William, who discovered Uranus in 1781, the first recorded discovery of a planet in human history). His only known comment is in a letter to his Aunt Caroline: 'I have been pestered from all quarters with that ridiculous hoax about the Moon – in English, French and German.'

The author of the articles, the British-born journalist Richard Adams Locke, later claimed that the story was a satirical piece written to show the gullibility of Americans on the question of extraterrestrial life.

56

The day of the self-replicating nanobots

NANOROBOTS OR NANOBOTS are hypothetical, intelligent machines too small to be seen by the naked eye. They were envisioned by Eric Drexler in his 1986 book, *The Engines of Creation*. Drexler, who is sometimes called the father of nanotechnology, imagined these microscopic robots as self-replicating. Like biological cells, they would be able to make copies of themselves. In theory, they could build anything as long as they had a ready supply of the right kinds of atoms, a set of instructions and a source of energy.

Richard Smalley, who won the 1996 Nobel Prize in chemistry, dismisses the idea, saying that 'self-replicating, mechanical nanobots are simply not possible in our world'. To achieve their aim, nanotechnologists must provide these nanobots with 'magic fingers'. Within the constraints of a space of one nanometre – the size of a nanobot – manipulation of atoms is not easy, because the fingers of a manipulator arm must themselves be made out of atoms. 'There just isn't enough room in the nanometre-size reaction region to accommodate all the fingers necessary to have complete control of the chemistry,' he says.

Let's apply this 'futurist's dream', as Smalley calls it, to intelligent aliens (*story 41*) – aliens capable of building a technological civilisation. Could they be so small that they are almost invisible? By Smalley's logic the answer is a definite no. They won't be able to build anything. However, on a large planet where the pull of gravity is strong, it is possible that any life form may not be very tall. Therefore, aliens of the size of a large insect are possible. They could possibly have highly intelligent brains. With their 'magic fingers' they could build small robots that could build still larger robots and so on, and thus could have a technological civilisation far more advanced than our own.

Though some primitive nanomachines have been developed, nanobots are still a hypothetical concept, and self-replicating nanobots a fiction.

57
Near death, not near god

WHEN OUR HEART stops beating and there is no blood flowing to our brain, all brain activity ceases and we are declared clinically dead. Is death instantaneous? No, doctors say that we take from a few seconds to a few minutes to go through the dying process.

What happens then? Obviously nothing as death is the final frontier and we have simply ceased to exist.

People who have come very close to clinical death and have survived do not agree. They tell amazingly similar stories about their experiences. The first stage of their 'life after life' is the feeling of immense peacefulness and the absence of pain and fear. They then somehow leave their physical body and find themselves looking down upon it. They continue to rise above their body and enter into a dark tunnel. Their peaceful journey ends when they see a light at the end of the tunnel. The distant, golden light is welcoming and some regard it as a supernatural presence of some sort. They think they have reached the boundary between life and afterlife. Some even recall speaking to their dead relatives or an encounter with certain aspects of their lives. What follows is the realisation that they have to leave this afterlife and then they wake up.

The details of such experiences may also have a religious aspect; for example, Christians tend to see Jesus in the light, while Hindus see Yamdoot, the messenger of death. But the core characteristics of the experiences are the same across all cultures.

People who have returned from the brink of death often see their near-death experiences as paranormal experiences. There is a soul or psyche dwelling in the physical body. When we die, this immaterial essence leaves the body and travels to another world.

Naturally, scientists disagree. They look for explanations in the brain, not in the supernatural world. They say that the brain surges with activity just before death. The retrospective analysis of brain activity of critically ill patients as they were removed from life support has shown that there is a significant spike in brain activity at or near the time of death. This increase in brain activity, which lasts from 30 to 180 seconds, could explain near-death experiences.

Some researchers associate near-death experience with the temporal lobes, the part of the brain that lies around ears. Studies suggest that electrical activity in the right temporal lobe is involved in mystical and religious experiences. Investigations of temporal lobe activity in people who had near-death experiences during life-threatening events have revealed that such a people have more temporal lobe activity than normal people.

Scientists have yet to fully explain the cause of near-death experiences, but it does not mean that these experiences are supernatural.

Nocebo effect

58

The evil side of placebos

DESCRIBED AS PLACEBO'S evil twin, the nocebo effect occurs when people claim to feel worse after taking placebo pills, which are inactive or dummy pills. Their headaches, fatigue, insomnia, stomach aches, nausea, dizziness and other symptoms are not in the mind; they are physical effects and can have long-lasting impact on health. Studies such as the Turin study described in placebo effect (*story 70*) show that a patient who expects to suffer painful symptoms is more likely to. When people think they are sick, they get sick.

Researchers noticed the first large-scale nocebo (from Latin *nocere*, 'to harm') effect in the late 1990s when they came across an unusual finding: Women, with similar risk factors, were four times more likely to die if they believed they were prone to heart disease. The higher risk of death had no underlying medical cause such as age, blood pressure, cholesterol or weight.

It seems that patients can create their own nocebo effect unwittingly. Women report nocebo responses to therapy more than men do but, like the placebo effect, the nocebo effect is not clearly related to individual variables such as gender, age and culture. Daniel Moerman, an anthropologist at the University of Michigan, has come across an interesting phenomenon about placebo pills. The most important thing about a pill is its colour. In Italy, he says, blue placebo pills made excellent sleeping pills for women but had the opposite effect on men. He found the answer to this puzzle in the Italian national football team's colour: blue.

Brian Olshansky of the University of Iowa Hospitals believes that a doctor may be an unwitting contributor to placebo and nocebo responses. 'A cold, uncaring, disinterested and emotionless physician will encourage a nocebo response,' he says. 'In contrast, a caring, empathetic, physician fosters trust, strengthens beneficent patient expectations, and elicits a strong placebo response.'

For most doctors, health is a biological phenomenon, they find it hard to deal with psychosomatic elements. To them, the nocebo effect is still a mystery, but you can defeat it by being aware of it. It's simply the matter of mind over mind.

N-rays

59
Delusion, blunder or hoax?

IN 1903 RENÉ Blondlot, a highly respected physicist at the University of Nancy in France, was experimenting with newly discovered X-rays when he noticed some strange things about the radiation coming out of his apparatus. He thought he had discovered a new kind of radiation and named it N-rays after his university. He claimed mysterious properties for N-rays, such as they could be stored in various things, a brick for example, and if you held the brick close to your head, the rays would increase your ability to see in the dark.

Blondlot was celebrated for his remarkable discovery and other physicists, mostly French, rushed to study this new phenomenon. They made more extravagant claims for these invisible rays: the rays were emitted by the Sun, flames and incandescent sources and any material in which strain was present such as hardened steel or brass.

Many scientists outside France were sceptical of these claims. In 1904, when American physicist Robert W. Wood learned about the N-rays, he decided to pay Blondlot a visit, while holidays with his family in Europe. Blondlot showed him the apparatus in his darkened laboratory and read various measurements loudly. Wood quietly put an aluminium prism, which was a crucial part of the apparatus, in his pocket and asked Blondlot to repeat the measurements. The new measurements were perfectly similar to the earlier ones. Wood published his report in the prestigious journal *Nature* claiming that N-rays were nothing but a delusion. 'What a spectacle for French science when one of its distinguished savants measures the positions of the spectrum line, while the prism reposes in the pocket of his American colleague!' exclaimed a French scientist.

Wood's report ended one of the biggest blunders in the history of science. Was it a case of self-delusion or a conscious effort to perpetrate a hoax? Whatever it was it did fool many respected physicists.

Numerology

60

Superstition by numbers

THE BELIEF IN lucky and unlucky numbers is probably one of the oldest superstitions. The sixth-century BC Greek mathematician Pythagoras (known for his theorem) played an important role in assigning magical qualities to numbers. 'Number is the ruler of forms and ideas, and it is cause of gods and demons,' he said. His lectures were so inspiring that some of his pupils formed themselves into a brotherhood for continuing his ideas. They called themselves Pythagoreans.

Pythagoreans considered that numbers ruled the universe. To them every number appeared to have been endowed with its own peculiar quality and character. They considered even numbers feminine and odd numbers masculine. They associated the number 1 with reason, 2 with opinion, 3 with mystery, 4 with justice, 5 with marriage, and so on. Relics of these fanciful ideas are still present in our language. For example, four dots form a square, and so a 'square deal' means justice. The number philosophy of Pythagoreans was later divided into two streams – number theory and numerology. As astrology (*story 9*) has no relation with modern astronomy, similarly numerology has no nothing to do with modern number theories. Numerology – a system that uses names and birth dates to reveal character traits and predict the future – is based on absolutely fanciful ideas, and the characteristics attributed to numbers are both arbitrary and personal.

To some Friday the 13th is the confluence of the unluckiest of days and the unluckiest of numbers. Psychologists will tell you that if you fear the Friday the 13th you may suffer from paraskevidekatriaphobia. The symptoms of this phobia range from mild anxiety to panic attacks. There is always at least one Friday the 13th during the year and never more than three. A comforting fact if you suffer from paraskevidekatriaphobia.

If you do not suffer from paraskevidekatriaphobia or any other number phobia and still believe in numerology, find your lucky number (there are numerous numerology websites to help you) and then go to a casino and bet all your money on your lucky number. You will soon find out how lucky your lucky number is.

The Egyptian 'heaven on earth'

IN 1979, AT London's Heathrow airport, Robert Bauval, a Belgian construction engineer, bought a book called *The Sirius Mystery* by Robert Temple (*story 82*). He was so fascinated by the book that he devoured it during his flight to Sudan. Though he found Temple's theory of fish gods of Sirius highly speculative, he decided to study the work of the French anthropologist Marcel Griaule, the foundation stone of Temple's speculations. The result of Bauval's labour was: *The Orion Mystery: Unlocking the Secrets of the Pyramids*, co-authored with Adrian Gilbert and published in 1994.

The constellation of Orion (the Hunter) is one of the most recognisable patterns of stars in the northern hemisphere. It has three bright stars in a row, which are known as the belt of Orion. One night in 1985, while he was camping in the Saudi desert, Bauval suddenly woke up. From his sleeping bag, he scanned the sky and saw the stars of the belt of Orion. It immediately dawned on him that the three largest pyramids of Giza were built as a giant map of the belt of Orion. Later, when he calculated the positions of the stars when pyramids were built, it showed the religious significance of the four shafts leading upwards from the heart of the Great Pyramid. He says that the southern shaft of the King's chamber pointed towards the belt of Orion, associated with the god Osiris; and the equivalent shaft from the Queen's chamber pointed towards Sirius, the star of the goddess Isis, wife of Osiris. Bauval says that, in fact, by building the pyramids of Giza, Egyptians wanted to create 'heaven on earth.'

Bauval has also tried to fit other pyramids of Egypt to complete the picture of the constellation of Orion. In order to make the pyramids to match the sky, you have to turn the map of Egypt upside down and even then, the match seems inaccurate. Egyptologists have dismissed *The Sirius Mystery* as a pseudoscientific pyramidology.

62

Out of body or out of mind?

OUT-OF-BODY EXPERIENCES – when a person sees his or her body from a location outside the physical body – are fairly common and researchers estimate that about 15 to 20 per cent of the population has had at least one out-of-body experience during their lives. People usually have such experiences when they are awake, not during sleep. The experiences can be spontaneous, but most often they occur during periods of high stress, serious illness, acute trauma, abuse of hallucinogenic drugs, hypnosis, meditation or religious prayers. There is no evidence that they are caused by mental illness.

Most out-of-body experiences have many common features. Persons having the experience observe their physical body lying motionless and are able to move around it. They perceive that they have another body, a double of the physical body; this ghostly or transparent body can pass through walls and other solid objects. Sometimes they see that their 'astral' body is connected to their physical body by a silvery 'umbilical' cord. As in near-death experiences, some experiences begin with the entry into a dark tunnel which has a bright, white light at the end of it. Most experiences are brief and end with a sense of pleasure and liberation.

Scientists do not doubt that people do have out-of-body experiences, but they do not agree that experiences are psychic, paranormal or mystical. Some neuroscientists have tried to induce elements of illusory out-of-body experiences in healthy volunteers by using electrodes to turn off the temporoparietal region of the brain (a region of the brain where temporal and parietal lobes meet; people with damage to this region are no longer able to map body's position in space and, for example, have difficulty negotiating their way around a house). These laboratory experiments have shown that perceptual illusion can be induced in which volunteers experience that their centre of awareness, or 'self', is located outside their bodies and they look at their bodies from the perspective of another person. These and other experiments suggest that consciousness of a self in our body is based on the processing of various sensory inputs in the brain, and the brain's representation of the physical body is changeable and can be modified by information from the senses.

The idea that soul or spirit can leave the body is an ancient one and is found in many cultures. For many, the most appealing explanation for out-of-body experiences is that the mind or soul, in fact, leaves the body. This explanation raises the question of how nonphysical mind could observe and reflect when it has left the physical body. Our mind depends upon the functioning of the brain in our body.

Are out-of-body experiences a purely psychological phenomenon, involving no soul or self leaving the body; or, are they a combination of imagination and extrasensory perception? Science has yet to provide a definitive explanation for this weird phenomenon but all the evidence shows them to be the misfiring of neurons, not paranormal or supernatural phenomena.

63

Your future world in the palm of your hand

PALMISTRY OR CHIROMANCY attempts to reveal character and foretell the fortune by reading lines, marks and patterns on the palm of the hand. The practice probably originated in ancient India and then spread to Greece, hence the name chiromancy from the Greek word for hand, *cheir*. The most popular form of palmistry is based on the Greek system.

Is there some truth in palmistry or is it simply hocus-pocus? A study, reported in 1974 in the *Journal of American Medical Association*, concluded that palmistry may be used to predict life expectancy, but when so used it is free of scientific worthiness or usefulness to life insurers. The researchers' conclusion was based on the examination of fifty-one cadavers and correlation of age at death with the length of life line (one of the three major lines on the palm, the other two are heart line and health line). A broken life line is not related to age at death and it is our personal experience that it correlates with nothing whatsoever, the researchers claimed. Other studies are equally scathing.

Hold up your right hand, palm towards you and finger together, and compare the length of the index (second) finger and the ring (fourth) finger. In general, women tend to have longer index fingers and men longer ring fingers. Research studies have associated longer index fingers with good verbal and literacy skills, where women dominate, and longer ring fingers with success in competitive sports such as soccer and basketball, where men dominate.

Whether you have a longer index finger or a longer ring finger, your palm is not going to reveal much about your future.

Panspermia

64

Life from outer space

SVANTE ARRHENIUS, THE Swedish chemist who gave us the chemistry of ions, was a versatile genius and investigated many scientific ideas beyond chemistry. In 1896 he was the first to recognise that carbon dioxide acts as a thermal blanket around the globe, thus creating the greenhouse effect. In 1906 he suggested that bacterial spores and other dormant microorganisms escaped from another planet where life already existed, travelled through space and finally landed on Earth and began to grow and develop. He called this process *panspermia* (Greek for 'all seeds'), but did not explain how life originated on other planets. He just said that life is eternal. It has always been there, so the question of its origin does not arise.

The idea of panspermia fascinated scientists of the 19th century; however, it never became an accepted scientific idea. In 1980 Fred Hoyle, one of the most outstanding scientists of the 20th century, revived the idea when he suggested that interstellar space contains organic molecules.

Scientists ridiculed the idea because the common belief then was that there were no molecules in space except atomic hydrogen and ionic hydrogen. However, in recent years many organic molecules such as benzene, sugars and ethanol have been detected in interstellar space. Amino acids, the building blocks of life, still elude scientists.

In 1981 Hoyle and another British astronomer Chandra Wickramasinghe suggested that life with all its basic genetic information originated not on Earth but on a grand cosmic scale. Chemical building blocks of life are present in interstellar clouds. When these clouds collapse to form comets, they provide likely sites for the origin of life. Microorganisms multiply inside a comet which has a warm, liquid interior. An impact of a comet about 3.8 billion years ago could have led to the start of terrestrial life. They said that only the minutest fraction (less than one part in a trillion) of the interstellar bacteria needs to retain viability for panspermia to hold sway.

Like comets, meteorites can also distribute microorganisms. Even today, about 100 tonnes of debris from comets and meteorites arrives on Earth daily. This cosmic debris may bring in to the planet microbes responsible for diseases of planets and animals. 'The boldest answer must be yes; that is to say extraterrestrial biological invasions never stopped and continue today,' Hoyle and Wickramasinghe say. They hold cosmic debris responsible for many epidemics of a global nature, like the influenza of 1918 and the plague of Athens in 430 BC. They point out many anomalies in the distribution and spread of the 1918 and 1968 flu epidemics, and conclude that simple person-to-person spreading was not an adequate explanation, whereas the atmospheric dispersion of a space-borne agent was more convincing.

All this sounds plausible, but the question remains how bugs can survive the lethal radiation in space. The answer probably is that in interstellar clouds a thin layer of carbon material forms around micro-organisms, protecting them against damaging radiation.

Recent research suggests that a galactic hitchhike by microorganisms is full of hazards but still feasible. Calculations by H. Jay Melosh of the University of Arizona show that microorganisms could even survive for millions of years in space if they are embedded in the interior of huge chunks of rock. This could happen when an asteroid impact ejects rocks from a planet. It seems that the impacts that produced craters on Earth greater than 100 kilometres across would each have ejected millions of tonnes of rocks carrying microorganisms into interplanetary space, much of it in the form of boulders large enough to shield those microorganisms from radiation. 'Although terrestrial organisms in these rocks would have the opportunity to colonise the planet,' Melosh says, 'it seems unlikely they would find the conditions suitable for propagation.'

It also seems unlikely that scientists will soon accept Hoyle and Wickramasinghe's idea of panspermia.

Paradigm

65

A hopelessly overused and abused word

HOW DOES SCIENCE progress? The eminent philosopher of science Thomas Kuhn answered this question in his seminal book, *The Structure of Scientific Revolutions* (1962). As a graduate physics student, Thomas Kuhn read Aristotle and Newton's works and realised how different were their concepts of matter and motion. This and other similar observations in the history of science – where there was a fundamental change in paradigm or the framework of thought–led him to the idea that science doesn't develop by the orderly accumulation of facts and theories, but by the 'tradition-shattering complements to the tradition-bound activity of 'normal science'. He called these shifts in paradigms scientific revolution.

To Kuhn, normal science – 'research firmly based upon one or more past scientific achievements, achievements that some particular scientific community acknowledges for a time as supplying the foundation for its further practice' – is simply 'puzzle-solving'. Revolutionary science, on the other hand, involves a complete revision of existing scientific beliefs and practices. The development of science isn't uniform, but has alternating 'normal' and 'revolutionary' phases. The worlds before and after a paradigm shift are absolutely different.

The Structure of Scientific Revolutions is one of the most popular and influential academic books of the 20th century. The book turned the words 'paradigm' (a typical example or pattern of something) and 'paradigm shift' (a fundamental change in underlying assumptions) into buzzwords. Even Kuhn (who died in 1996) complained that the word 'paradigm' had become 'hopelessly overused' and was 'out of control'. The *New Yorker* magazine was inspired to publish a cartoon in which a woman exclaims: 'Dynamite, Mr. Gerston! You're the first person I ever heard use "paradigm" in real life.' If Google is any guide to the popularity of a word or phrase, note the following statistics: In 1997 it listed 42 million pages for 'paradigm', of which 2.6 million were 'paradigm shifts'. In 2010 it listed staggering 85 million pages for 'paradigm', of which 8 million were 'paradigm shifts' (in real life, we hope).

Most of the paradigm shifts we hear or read about these days are not really paradigm shifts. In a 1964 review the prestigious *Scientific American* magazine dismissed Kuhn's book as 'much ado about very little'; however, the book was indeed a paradigm shift in the philosophy of science. It has greatly influenced the way philosophers, historians and sociologist see scientific change, but Kuhn's original thesis is now generally seen as limited.

Not everyone agrees with the idea that science is not a continual building process. It is difficult to accept that the history of science is not clearly characterised by a natural progression in scientific change. Some critics argue that a close look at historical episodes fails to identify Kuhn's stages of scientific change. Others say that paradigms might help in clearly defined problems, but they tend to move scientists away from pressing problems that cannot be neatly argued.

Parallel universes

66

They have found gazillion copies of you

THERE ARE COUNTLESS universes and in each of them a copy of you is reading this article right now. The idea of parallel universes has launched a thousand science-fiction stories, but it's no longer the staple of science fiction. Many scientific theories now support this bizarre idea.

Scientists agree that the universe (the one in which real you are reading this article) began with a Big Bang about 13.7 billion years ago when an unimaginably dense and unimaginably hot speck of matter exploded spontaneously. The newly born universe was opaque as there was no light. After 380,000 years the universe cooled to about 4,500°C (8132 °F) and it became transparent as photons, the particles of light, were free to escape. This fossil light has now turned into microwaves and fills the universe. Referred to as the cosmic microwave background, it can be detected anywhere and has a temperature of −270°C or −480°F (3°C or 37°F above absolute zero).

Spacetime, the fabric of the cosmos, could have only three possible kinds of surface: curved like a giant ball, warped like an enormous saddle, or flat. The cosmic background radiation – its existence was confirmed experimentally in 1965 – shows no distortion.

This means that the universe is flat; a curved or warped universe would distort the size of distant objects such as the remnant cosmic background radiation. Other cosmological observations also support the idea of a flat universe.

A flat universe suggests that space is infinite in size. This infinite space holds infinite copies of our universe. All these universes are part of a larger multiverse. As a citizen of the multiverse, could we travel to other universes? Gravity, the warping of spacetime, can freely float into the space between universes, but gravity also keeps us glued to our universe and stops us leaping off it into another universe. We cannot even see our other selves. The farthest we can 'see' is the distance light has travelled since the Big Bang. The actual distance we can currently observe is much smaller.

Perpetual motion machines

67

Running forever without energy

BHASKARA (ALSO KNOWN as Bhaskaracharya, 'Bhaskara the Learned') was a mathematician who lived in India in the 12th century. He has a respected place in the history of mathematics. He wrote the first works using the decimal number system which described rules for calculating with zero and the concept of negative and positive numbers. He also has a special place in the annals of perpetual motion machines as the designer of a wheel that could turn forever, although he never built it. The wheel, with containers of mercury around its rim, was designed to rotate constantly, because the wheel would always be heavier on one side of the axle. We know now that such a machine cannot be built.

This idea, like the ideas of zero and decimal numbers, re-appeared in Arabic writings. From the Islamic world it reached the Western world. In the 15th century even Leonardo da Vinci sketched many designs in his notebooks. But the great English scientist William Gilbert was scornful of the claims of magnetic perpetual motion machines. In 1600 he published *De Magnete*, the first ever learned work in experimental physics in which he proved that Earth is a magnet and showed that an iron rod can be magnetised by forging it in the north-south direction. His science was on the ball when he criticised – with the full force of invectives – exclaiming, 'May the gods damn all such sham, pilfered, distorted works, which do but muddle the minds of students.'

Nevertheless, over the centuries numerous scientists and engineers tried – and failed miserably – to build perpetual motion machines. Failure did not stop them patenting their creations. The first patent for a perpetual motion machine was granted in 1635 in England. Most countries now refuse to grant patents for perpetual motion machines without working models, but they can still be patented in the United States and Canada as these countries do not require a working model.

A perpetual motion machine has to remain in motion forever without applying any force or consuming any energy. This idea violates the sacrosanct laws of thermodynamics. The first law of thermodynamics (which is the law of conservation of energy) demands that no machine can produce more energy than it uses.

The second law of thermodynamics places constraints on machines, such as car engines, that do useful work by tapping heat energy from burning fuel. This law requires that heat must flow from a hotter to a colder body. This means the machine must lose some energy when heat is converted into useful work.

68
Divining character from bumps on the head

IN 1798, FRENZ Gall, an Austrian physician, published a paper in a German journal in which he claimed 'that it is possible by observing various elevations and depressions on the surface of the head to determine the degrees of different aspects of the personality.'

This 10-page article marked the beginning of phrenology (*phrenos* is Greek for mind), the study of the shape and size of the head to determine a person's character and mental abilities.

Gall made two major assertions. First, he believed that different mental functions are located in different parts of the brain, called organs. Second, he argued that the growth of the various organs is related to the development of associated mental faculties. As this growth would be reflected in the shape of the skull, personality traits could be determined by reading bumps or indentations on the skull.

He even boasted that phrenology will be of the first importance to medicine, morality, education and the law – indeed to the whole science of human nature.

Gall identified 27 discrete organs of behaviour (*see accompanying picture*); this number increased over time as new organs were 'discovered'. If you move your finger on the back of your neck, you will notice a bump formed by the base of your skull. This bump, according to Gall, marked the location of the organ of Amativeness, the organ that defined the attachment of sexes to each other. If a person's head showed a comparatively large organ of Amativeness, the person was sexually unrestrained. A comparatively small organ of Amativeness showed indifference towards the opposite sex.

Phrenology is, of course, quackery, but in the early 19th century it was considered a respected science. People sought the advice of phrenologists not only for diagnosing mental illness, but for hiring employees, or even selecting marriage partners.

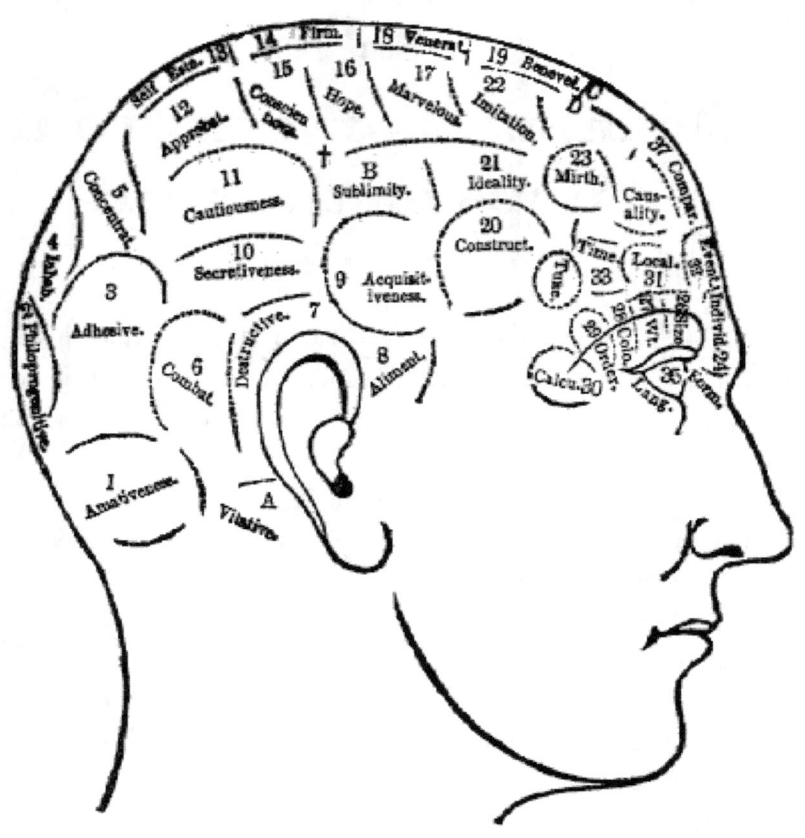

Piltdown man

69
A synonym for phoney science

A SKULL – A blend of human and ape – was discovered in 1912 in a gravel pit in Piltdown, a village in Sussex, England. It was described as a 500,000-year-old fossil and the proof of humans' ape-like ancestry – the 'missing link' between apes and humans that anthropologists have been searching for since Darwin postulated the existence of intermediate forms between ancestral apes and modern humans in 1871 in his book, *The Descent of Man*.

Piltdown man was hailed by many British anthropologists as the most ancient ancestor of modern humans. It was named *Eoanthropus dawsoni*, 'Dawson's dawn man', after Charles Dawson, a lawyer and a keen amateur archaeologist, who found the remains. Proudly called 'the first Englishman', as it gave England an important place in the history of humankind, the Piltdown man was cited along with Neanderthal man and Heidelberg man, both discovered in Germany, as an example of early hominid life in Europe. When fossils of *Australopithecus*, a primitive hominid, were found in Africa in the 1930s, Piltdown man became an enigma. In 1953 chemical analysis and radiocarbon dating showed that the fossil was a fraud: the lower jaw was that of a female orangutan deliberately coloured to look old; the cranium was of human origin and was less than 1,000 years old.

Piltdown man is one of the greatest forgeries in the history of science, which fooled many of the best minds in science for four decades. Thousands of textbooks had to be revised when this fake ancestor of humans was unmasked. The perpetrators of the fraud have never been identified. However, various people have come under suspicion, including Dawson (who died in 1916), Arthur Keith, an eminent anthropologist and anatomist who vigorously supported the idea that cranium and lower jaw belonged to the same skull and the fossil was indeed the 'missing link', and even Arthur Conan Doyle, the creator of Sherlock Holmes, who lived in Sussex and played golf at Piltdown.

Placebo effect

70
Is belief one of the most powerful medicines?

A TYPICAL DRUG trial goes something like this. Researchers randomly divide the patients in three groups: the first group is given the drug being tested, the second gets no treatment and the third the placebo. The placebo (from Latin *placere*, 'to please') is typically a pill that looks and tastes like the drug but doesn't contain any drug. The no-treatment and placebo groups are known as control groups, the former shows how many patients would likely to get better by themselves and the latter shows the effect of belief in the drug.

The results are always surprising: the placebo group may have zero to 100 per cent recovery rate, and the results are not clearly related to individual variables such as gender, age and culture (surprisingly, some trials have shown variation in results from country to country). Researchers say that the placebo effect depends on too many variables to have a simple relationship with any one of them.

However, placebo treatment works better if the doctor treating you believes in it. A study carried out in 2003 by neuroscientist Fabrizio Benedetti of the University of Turin in Italy showed how the doctor's words affect the patient's brain. His team induced severe arm pain in a group of healthy volunteers. To control the pain the participants were then given a saline injection. Over the course of four days, some participants were told that the saline injection was a powerful painkiller (a verbal suggestion aimed at inducing expectation of analgesia), others that it increased pain (expectations of hyperalgesia), and the third group wasn't told anything. The results showed that anticipating less pain led to a significant decrease in pain, while anticipating more pain led to increased agony. The belief of the doctor and his or her verbal suggestions to the patient (this medicine is good for you) increases the efficiency of placebo treatment. Perhaps this is the reason why alternative therapies work; their believers have great faith in their therapy and they are also able to convince their patients of the effectiveness of their treatment.

Until now medical researchers have considered the placebo effect a nuisance when testing a new treatment, but their view is changing. They believe that a patient's expectations and beliefs can greatly change the course of an illness, and see placebos as a key to understanding how the brain promotes faster healing. Intriguingly, you can experience a placebo effect even if you don't believe in it. Does the subconscious play a part? There is a lot we don't know about placebos. Is this bogus treatment real? Reserve your judgment.

Planet Nibiru

71
A fictional planet to end the world

DID YOU KNOW that there is a rogue planet orbiting the Sun every 3,600 years? This planet is called Nibiru and on 21 December 2012 it was expected to enter the solar system and crash into Earth. On that day, cataclysmic events were expected to occur according to the Mayan calendar (*story 51*). Obviously, nothing happened. And, if you believed in conspiracy theories, NASA was tracking this planet but keeping it secret as a part of worldwide conspiracy.

How did a planet never recorded by astronomers and space scientists become so popular that millions and millions of websites were devoted to it? The story of Nibiru began in 1976 when Zecharia Sitchin claimed in his book, *The Twelfth Planet*, that there is an undiscovered planet beyond Neptune. He said that he has found and translated ancient Sumerian texts that reveal that Earth was struck by a large planet called Nibiru. The collision moved Earth into its present orbit and created the Moon and the asteroid belt.

Sumerians, an urban civilisation that flourished on the banks of Tigris and Euphrates rivers in Mesopotamia (now Iraq) from about 23rd century BC to 17th century BC, was indeed a great ancient civilisation but they have left very few records about astronomy. According to David Morrison, a senior NASA scientist, Sumerians were not even familiar with the idea that planets orbited the Sun, an idea that first appeared in ancient Greece two thousand years after the end of Sumer. Sumerians didn't know about the existence of Uranus, Neptune or Pluto. Morrison disputes the notion that they even had a god named Nibiru. The name appears in a Babylonian creation poem Enuma Elish recorded in seventh century BC, centuries after the Sumerians. It's associated with the god Marduk, generally accepted as referring to planet Jupiter. He stresses that the rest of the story is the product of Sitchin's imagination.

Like Erich von Daniken (*story 5*) and Immanuel Velikovsky (*story 21*), Sitchin assumes that ancient stories are not myths but scientific facts. The simple fact is that the planet Nibiru doesn't exist. Originally, the planet Nibiru was predicted to hit Earth in May 2003. Then doomsayers reset the date to 21 December 2012 to coincide with other end-of-the-world scenarios such Pole reversal and pole shift (*story 73*). The focus and dates of doomsday fear keeps on shifting. Why worry about junk science when you can worry about real science such as climate change?

Doomsayers sometimes confuse Nibiru with Planet X (*next story*), the generic name given to any suspected or possible planet beyond Pluto, or Eris, a newly discovered dwarf planet.

Planet X

72

The saga of an unknown planet

THE ANCIENTS KNEW of only five planets – Mercury, Venus, Mars, Jupiter and Saturn, all named after Roman gods. Until 1543, when the Polish astronomer Nicolaus Copernicus declared that 'Earth was simply one wanderer of many', Earth was believed to be motionless and not a planet.

The discovery of Uranus in 1781 posed a problem for astronomers. They kept on finding it in the wrong parts of the sky; the planet was apparently drifting away from its predicted orbit. Was there an unknown planet beyond Uranus pulling it out of its orbit?

Even after Neptune's discovery in 1846, Uranus kept on misbehaving. In 1895 Percival Lowell, a wealthy American astronomer, suggested that an undiscovered planet was affecting Uranus and Neptune. He called it Planet X (X as in unknown). He started searching for it from the Lowell Observatory near Flagstaff in Arizona, which he had built, partly for the very purpose of finding Planet X. He spent 10 years searching, but found nothing. A disappointed man, he died in 1916. However, his observatory continues to operate.

After the discovery of Pluto in 1930, astronomers asked 'is it Planet X?' No. Pluto is so tiny that it has no appreciable effect on the motion of either Uranus or Neptune. In the 1990s astronomers recalculated the masses of Jupiter, Saturn, Uranus and Neptune, using data from Voyager fly-bys, and concluded that Uranus and Neptune were right on course. There is no Planet X affecting Uranus and Neptune.

Pluto has now been relegated to a new class of objects called dwarf planets (there are only eight planets in the 'new' solar system). This new category also includes Ceres, discovered in 1801. In 2003 American astronomers discovered an object which they thought was Planet X. They informally called it Xena as a word play on Planet X (in 2006 it was officially named Eris). Like Pluto, Eris is a rocky ball covered with frozen methane. A little larger than Pluto, it lies about three times as far from the Sun as Pluto, which makes it the furthest object ever seen in the solar system. It has also been classified as a dwarf planet.

But the myth of Planet X refuses to die. Conspiracy theorists say that the orbit of this mysterious and rogue planet is coming closer to Earth and causing extreme weather and major earthquakes. They also blame it for an increase in solar flares erupting from the surface of the Sun.

Solar flares rise and fade slowly, over years. When a solar flare peaks it unleashes waves of energy, disrupting Earth's magnetic field, which in turn can damage power grids and communication satellites. Our dependence on technology places us at a greater risk if power grids and satellites stop working.

Pole reversal and pole shift

73

Not the end of the world

EARTH HAS A very strong magnetic field that extends some 60,000 kilometres (37,280 miles) out in space. You can imagine this magnetic field as a big bar magnet inside Earth. It has north and south poles and is slowly moving. At present, Earth's geographical north and south poles are not pointing in the same direction as magnetic poles. There is a difference of about 11 degrees.

When certain rocks are formed, small grains of iron act like tiny compasses and line up in the direction of Earth's magnetic field. When the rocks solidify, these little 'compasses' are locked in. The magnetic field is 'fossilised' in the rocks. The study of these rocks shows that the magnetic poles have reversed, or flipped by 180 degrees, many times in the past. In the last 10 million years there have been, on average, four or five reversals per million years. The last one was about 780,000 years ago when compass needles would have pointed south. Pole reversals are the result of movement of molten iron in Earth's outer core.

Scientists have observed that Earth's magnetic field has weakened in a certain location over the Atlantic, off the coast of Brazil. This unusual distortion in the magnetic field, known as South Atlantic Anomaly, had prompted apocalyptic believers to link it with the Mayan calendar (*story 51*) that cataclysmic events would occur on 21 December 2012. Obviously, nothing happened on that day.

Though Earth's tear-shaped magnetic field shields us from deadly radiation such as cosmic rays, the atmosphere also acts as an extra blanket to stop most high-energy radiation. There is no obvious correlation between magnetic pole reversals and mass extinctions of the past. Some animals such as pigeons and whales do depend on Earth's magnetic field for navigation. Pole reversals do not happen overnight, they take tens of thousands of years. The future generations of these animals will certainly adapt to any changes in Earth's magnetic field.

If pole reversals won't end the world, pole shifts would, at least, according to doomsayers. Could a sudden pole shift spell doom for civilisation?

In 1842, Joseph Adhémar, a French mathematician, published a book in which he said that the melting and collapse of polar ice caps could cause Earth to flip by destabilising its centre of gravity. This idea was picked up by Charles Hapgood, an American professor of history, who in the 1950s published a book, *The Path of the Pole*, in which he suggested that increasing masses of ice, accumulated near a pole, could cause Earth's rigid crust to slip over its molten core. This would displace the polar region towards the equator. In other words, it would cause a pole shift, which would cause a global tsunami.

Hapgood even persuaded Einstein to write the foreword to his book. In his foreword, Einstein noted: 'Without doubt Earth's crust is strong enough not to give way proportionally as the ice is deposited.'

More recently, the idea of pole shift has earned support from Adam Maloof, a Geologist at Princeton University. He believes that pole shift is a fascinating and important process in geological history, but it has nothing to do with any apocalyptic scenario. The eruption of a large volcano in the polar regions, the impact of a meteorite or the melting of polar ice caps, he says, might cause a very small pole shift. Large pole shifts have happened in the past and such an event occurred 800 million years ago. But it had been at least a million years in the making.

You can safely bet your proverbial two bobs on no pole reversal and no pole shift happening in your lifetime.

74
Contaminated with silica and silliness

THE AMAZING STORY of polywater began when in 1962 N.N. Fedyakin, a Russian scientist, claimed the discovery of a water-like liquid, which he called 'anomalous water'. This gel-like substance was formed during condensation of water vapour in quartz capillary tubes. It was claimed that it was heavier and more viscous than water; it boiled at about 540°C (1,004 °F) and froze at −40°C (−40 °F) into a glassy substance quite unlike ice.

The discovery fooled hundreds of scientists worldwide who churned out a stream of research papers from 1962 to 1974 describing its incredible properties. Some respected United States scientists even hypothesised that it was a form of water in which the molecules are bound together in long chains, or polymers, like those in plastics, and named it polywater. Others worried that if released from the lab, it would propagate itself by feeding on natural water, turning our planet into another freezing Venus.

The bubble of scientific enthusiasm burst when carefully controlled analysis of minuscule samples provided by Fedyakin showed that it was contaminated badly by organic compounds. Polywater could only be prepared in quartz capillary tubes. Quartz is silica or silicon dioxide and is slightly soluble in water. Polywater was nothing but water contaminated with silica.

When polywater disappeared from the scientific world, it found a new home in *Star Trek*'s science fiction universe, where it belonged in the first place.

Why did polywater become a popular subject for research in spite of the fact that not many scientists had actually tested the substance? Most of the polywater frenzy was fuelled by the wide coverage of polywater stories by the media. When it became a popular subject, many scientists used the opportunity to attract the attention of the media and their peers at meetings. Even scientists like their fifteen minutes of fame. It wasn't science's finest hour. Score: silliness, 1; science,

Psychoanalysis

75

Still on the therapist's couch

YOU MAY HAVE never read a word of Sigmund Freud's major works – *The Interpretation of Dreams* (1900) and *Three Essays on the Theory of Sexuality* (1905) – but you probably know some Freud speak: Oedipus complex, id, ego, superego, sexual sublimation, repressed desires, and so on. And you can't say that you have never made a Freudian slip (an unintentional error in speech that Freud would have explained as a message from your unconscious mind revealing your suppressed thoughts or feelings).

Freudian psychoanalysis is based on the belief that our emotions and behaviour arise from unconscious fears and desires. The past shapes the present, and if we can trace the source of our unconscious fears and desires to their historical origins – often our childhood experiences – we can understand our troubles and deal better with the realities of life. All you have to do is to lie on the therapist's couch and talk about anything that comes to mind, and the source of your current problems would slowly begin to appear.

Freud's influence on Western thought has been extraordinary. He changed the way we see ourselves, but the question remains: is psychoanalysis an objective science or a pseudoscience? Psychoanalysis has not yet proved itself empirically a science; nor has psychoanalysis yet been proved quackery. Sceptics continue to challenge psychoanalysis, yet no better theory has yet emerged to replace it completely.

'If often he was wrong and, at times, absurd, to us he is no more a person now but a whole climate of opinion.' This comment from the poet W. H. Auden after Freud's death in 1939 still holds true, and the present climate of opinion is cloudy.

Psychokinesis

76

Mind over matter

PSYCHOKINESIS (ALSO CALLED telekinesis) is the alleged ability to move objects by mental effort alone. Simply put, it is mind over matter. Acknowledged in many cultures since the dawn of history, psychokinesis gained enormous publicity in the early 1970s when the Israeli psychic Uri Geller appeared on television shows around the world claiming to bend spoons and other metal objects apparently with the force of his thoughts. The Force wielded by the Jedi Knights in the popular *Star Wars* movies is also psychokinesis.

Geller's spoon-bending trick has been shown as a quick sleight of hand, not the result of psychic powers. The Force is simply science fiction. But no amount of demonstrable fraud or fiction can dissuade the true believers in the power of paranormal.

Psychokinesis is inconsistent with the laws of physics which have been proved beyond any reasonable doubt, creating problem for scientists who wants to study it. Another problem, says American theoretical physicist Michio Kaku, is that scientists are easily fooled by those claiming to have psychic power. 'Scientists are trained to believe what they see in the lab,' he says. 'Magicians claiming psychic powers, however, are trained to deceive others by fooling their visual senses. As a result, scientists have been poor observers of psychic phenomena.' Einstein put it in a subtle way: 'Nature hides her secrets through intrinsic grandeur but not through deception.'

These views didn't deter Dr Robert G. Jahn, a professor of engineering, from founding in 1979 the Princeton Engineering Anomalies Research laboratory at Princeton University. Until it was closed in 2007, the laboratory managed, according to *The New York Times*, 'embarrass university administrators, outrage Nobel laureates ... and make headlines around the world with its efforts to prove that thoughts can alter the course of events.'

The event most studied by Dr Jahn's laboratory was the flipping of coins. The theory of probability tells us that the when we flip a coin there is a 50 per cent chance of getting heads or tails. Dr Jahn's team designed an experiment which was an equivalent of a coin flipper. Volunteers would sit in front of an electronic box that flashed random numbers just above or just below 100, but had no physical connection to the machine. They would then try to influence the outcome by thinking 'high' or 'low' to produce a higher or a lower number than it should be by chance.

When the researchers looked for difference between the machine's output and random chance after more than two million 'coin flips' over two decades, they found the effects very small but amazing. The volunteers 'thoughts' were roughly altering one number in 1,000. In other words, if you had a coin flip, psychokinesis could affect one of those coins flips if you flipped a coin thousand times. Scientists dismiss this and other similar data on the ground that such minute differences could have been caused by subtle, hidden biases in their experiment's design. If you disagree, you might try psychokinesis in front of a poker (or slot) machine. Think positive and be nice to the machine and it will come up with the jackpot, 'You can do it, darling, sure you can.'

Of course, psychokinesis enthusiasts insist that it's possible, in varying degrees, for the human mind to influence the physical environment. But no one has ever produced a satisfactory theory to explain their claim. Some have turned to quantum mechanics to explain how mind might affect matter (*story 79*).

Pyramid power

A free way to sharpen your razor blades (if you use them)

IN OCTOBER 1931 Gilbert Coleridge, a London *Times* reader, wrote in a letter to the editor about his experiment on the magnetisation of safety razor blades: 'I conceived it possible that if I orientated my razor blades – i.e. kept them lying N. and S. by the compass, it might affect the life of the blade.' The experiment, repeated over seventy-two days, suggested that terrestrial magnetism did have the effect of lengthening the average life of a razor blade. 'Can any of your readers throw light on this question?' he asked. Well, one expects to pay attention to the question of a gentleman whose name links William Gilbert (the English scientist who in 1600 was the first to realise that Earth is itself a giant magnet) and Samuel Taylor Coleridge (the English poet best known for 'The Rime of the Ancient Mariner', 1798). As there was no spirited discussion of the science or poetry of the question in the letters pages, we can only conclude that *Times* readers were not energised by the idea of obtaining a longer-life blade by fondly watching it floating N–S on a cork in a basin of water.

The idea of sharpening razor blades without spending any energy appeared again in 1959 when the patent office in the then Czechoslovakia granted a patent to Karel Drbal, a radio engineer. His device was simple: a hollow pyramid, preferably a four-sided one with a square base, made of a non-conducting material such as thick paper, cardboard or plastic. The patent advised the blade be placed in N-S direction on a non-conducting material such as cork and then covered with the pyramid in such a way that pyramid's walls faced north, south, east and west. Drbal claimed that his invention could save both valuable material and energy. This was before tree-hugging became popular and his advice to save energy was ignored. But not for long.

In the early 1970s Patrick Flanagan, an American inventor, picked up Drabl's pyramid idea and said that ancient Egyptian pyramids and their scale models generated special kinds of energy. He coined the term 'pyramid power' and wrote several books on the subject.

Others also experimented on supernatural powers of pyramids and claimed that besides sharpening blades, models of pyramids could be used to preserve food, keep milk fresh, improve memory, calm children, heal cuts and cruises, increase psychic energy ... the list goes on. In the 1970s such claims helped some cardboard pyramid manufactures rake in millions of dollars.

In today's energy-starved world, any idea that can inspire new energy-saving technologies is worth serious consideration. It's a pity that no scientific experiment has ever succeeded in harnessing the power, natural or supernatural, of pyramids. In 2005 Discovery Channel's *Mythbusters* also staged a number of experiments to verify the claims of pyramid power proponents. Their conclusion: pyramid power is bogus.

The non-physics of holistic healing

ENERGY IN ELECTROMAGNETIC radiation is not a continuous quantity but flows in discrete 'packets' or quanta (singular: quantum). When particles emit energy, they do so only in quanta. This is the bare essence of quantum theory for non-physicists. The theory was proposed in 1900 by the German physicist Max Planck. The idea of quantum gave birth to the new quantum mechanics, which deals with the behaviour of nature at the atomic level. It suggested that the events at atomic level occur randomly. This spelled the doom of determinism and introduced the idea of uncertainty: it is impossible to find out exactly both the position and momentum of a particle. To measure both simultaneously requires two measurements: the act of performing the first measurement will 'disturb' the particle and so create uncertainty in the second measurement. This means that the mere act of observation changes the behaviour of the object. Does reality change when we try to observe it? Quantum mechanics is mind boggling. Even super-brilliant physicist Richard Feynman once admitted that it 'appears peculiar and mysterious to everyone – both to the novice and to the experienced physicists.'

What does a theory of matter has to do with mind and body? This anomaly has not stopped New Age gurus to draft 'quantum mechanics' in the service of their so-called holistic healing practices.

At the forefront of holistic-healing movement is Deepak Chopra, an endocrinologist, spiritual leader and author of many bestselling books, who came up with the idea of quantum healing: healing the body-mind from a quantum level. He maintains that 'quantum healing is the ability of one mode of consciousness (the mind) to spontaneously correct the mistakes of another mode of consciousness (the body)' and that illness and even the ageing process can be banished by the power of mind.

Chopra says that mind-body duality is the chief cause of all illness and we can cure all our ills by the application of sufficient mental power. He asserts that 'simply by localising your awareness on a source of pain, you can cause healing to begin, for the body naturally sends healing energy wherever attention is drawn.' Yes, studies have shown that there is some connection between mind and both illness and recovery, but can mind control serious illnesses such as cancer?

Quantum healing is a mish-mash of ideas collected from quantum mechanics, ancient Hindu metaphysics and Ayurveda, the ancient Hindu system of medicine. This quantum quackery lacks robust – even weak – evidence, and yet it appeals to millions of intelligent, educated people. Why? When people read in the popular media about speculative scientific ideas such as mirror matter, wormholes through spacetime, quantum teleportation, 11-dimensional superstrings and parallel universes, they tend to believe them without realising that science progresses by hypothesising and that not all hypotheses become accepted scientific ideas. The seductive idea of a universe full of mysteries makes people believe that everything is possible and we can achieve what Chopra calls 'ageless body, timeless mind' by the sheer force of consciousness.

Using the word 'quantum' in the context of healing would have made Planck cringe the same way it does modern physicists. Though his theory had predicted the existence of quanta or photons – a photon is an elementary particle representing a quantum of light or other electromagnetic radiation – he was doubtful about their reality. He once visited a physics lab where he saw in action an apparatus that counted photons by audible clicks. He stood silently for a while and just listened. Then he smiled and murmured, 'So they do exist.'

Does quantum healing exist? We must apply the same scepticism to quantum healing as Planck applied to his quantum theory.

Can quantum mechanics save your soul?

AT THE QUANTUM level, matter exhibits dual nature: particles sometimes behave as waves and sometimes as particles. For example, light, as an electromagnetic radiation, is transported in photons that are guided along their path by waves. This is called wave–particle duality: waves and particles are not two separate entities but the characteristics of the same substance. At a given moment, whether an entity is a particle or a wave depends upon what you measure. If you measure its position, you will conclude that it's a particle, but if you measure its wavelength you will conclude that it's a wave. Furthermore, measurement of one quantity will affect the value the other quantity will have in a future measurement. This creates uncertainty in measurement as the value to be obtained in a future measurement is unpredictable.

In quantum mechanics, we can only calculate probabilities, while in pre-quantum classical physics we calculated what actually happens. Einstein hated the probabilistic nature of quantum mechanics. He thought that the universe was deterministic in some ways. He was fond of saying, 'God does not play dice.' Once Niels Bohr, who proposed the quantum model of the atom, retorted, 'Who are you to tell God what to do?' And Stephen Hawking (famous for *A Brief History of Time*) has added his own wisdom, 'God sometimes throws dice where they can't be seen.'

The enigmatic concept of wave–particle duality concept has led to the idea that the true nature of the universe is not objective, but depends on the consciousness of the observer. This, in turn, has propagated the view that there are parallels between quantum mechanics and mysticism.

In 1975 Fritzof Capra, a professor at the University of California at Berkeley, explored 'the parallels between modern physics and Eastern mysticism' in his book, *The Tao of Physics*. The central thesis of Capra's interesting and highly successful book is that 'a consistent view of the world is beginning to emerge from modern physics which is harmonious with ancient Eastern wisdom'. His 'Eastern wisdom' is derived from the tenets of Buddhism, Hinduism and Taoism. And his 'consistent view' includes: (1) a holistic view of reality – there is no observer separate from reality and there is no separate reality from observer; (2) there is one ultimate reality which underlies and unifies the multiple things and events we observe; (3) paradoxes such as wave – particle duality are fundamental sources of insights; (4) mystics have an intuitive comprehension of the modern relativistic concept of spacetime; and (5) the equivalence of the quantum field and ch'i, the Chinese concept of natural energy of the universe.

Capra's book has made 'quantum' a buzzword which is now used to give scientific respectability to pseudoscientific spirituality of the New Age. Quantum mechanics is misinterpreted to imply that at some level our minds are in control of reality, or paranormal phenomena such as out-of-body experiences (*story 62*) are examples of the nonlocal operation of consciousness.

The world of quantum mechanics is as mysterious as ever, but it doesn't need Eastern mysticism to explain it. Belief is necessary for our survival but not quantum mysticism.

Bhagavad Gita, one of the sacred scriptures of Hinduism, describes three types of knowledge:

> When one sees Eternity in things that pass away and Infinity in finite things, then one has pure knowledge.
> But if one merely sees the diversity of things, with their divisions and limitations, then one has impure knowledge.
> And if one selfishly sees a thing as if it were everything, independent of the one and many, then one is in the darkness of ignorance.

The question is: Which one is quantum mysticism preaching?

80
Invisible and imprecise ink

IN 1921 HERMANN Rorschach, a Swiss psychiatrist, published a monograph, *Psychodiagnostik*, in which he presented a set of ten inkblots to be used for probing the personality. When he was at school, Rorschach was nicknamed Kleck (inkblot) by his friends because of his interest in making fanciful inkblot pictures.

Klecksography, placing an ink blot on paper and folding it in half to obtain the form of a butterfly or a bird, was a game popular among Swiss children in the 1920s. Rorschach died in 1922, at the age of 37, before he had barely begun to broaden the application of his test from mental patients to 'normal' persons. His test quickly became household cliché for psychological testing all over the world.

Rorschach's same ten cards are still in use today. Each has a symmetrical inkblot: five inkblots are black and grey, two are black and red, and three are multicoloured. An examiner hands one card at a time to the subject and asks what the subject sees, or thinks he or she sees. The subject is allowed to rotate the card to see it from different angles. The answers are said to be a 'projection' of the subject's personality. The ambiguous blots tell a different story to every person. Advocates of the test claim that what seem, superficially, to be chance associations actually reveal deep aspects of personality, and trained psychologists could use your interpretations of the blots to reach conclusions about your personality traits and emotional makeup.

Since 1950 many psychologists have dismissed the Rorschach test as 'subjective' and 'projective' in nature, merely a variation on Tarot card reading; yet the test is still administered by many psychologists, especially in the United States. After reviewing a large body of research on the Rorschach test American psychologists Scott O. Lilienfeld, James M. Wood and Howard N. Garb have come to the conclusion that 'it is poorly equipped to identify most psychiatric conditions – with the notable exceptions of schizophrenia and other disturbances marked by disordered thoughts ... the method does not consistently detect depression, anxiety disorders or psychopathic personality' (*Scientific*

American May 2001). They say that the test may be even more misleading for minorities, and the collected research raises serious doubts about its use in the psychotherapy office and the courtroom.

SEE THE FOLLOWING PICTURE: In 2009 Wikipedia published the ten actual Rorschach inkblots along with standard interpretations (en.wikipedia.org/wiki/Rorschach_test). The following inkblot is the second in the series. The main body of the image is in black and grey except the two 'heads' at the top and the two 'legs' at the bottom, which are in red. According to Wikipedia, three popular responses are: (1) two humans, (2) four-legged animal, and (3) animal: dog, elephant bear; and psychologists' common comments are: 'The red details are often seen as blood, and are the most distinctive features. Responses to them can provide indications about how a subject is likely to manage feelings of anger or physical harm. This card can induce a variety of sexual responses.'

Psychology Today magazine (November/December 2009) has published 'A faux diagnostic guide' to this inkblot. An excerpt: If you see 'an X-ray', 'you may be a psychoanalyst', and if you see 'tea leaves', 'you're analyzing the analysts, thinking they project what they want to see onto the results … a few of them have used palm-reading techniques to fool others, and many of them are genuinely mistaken about the test's validity.'

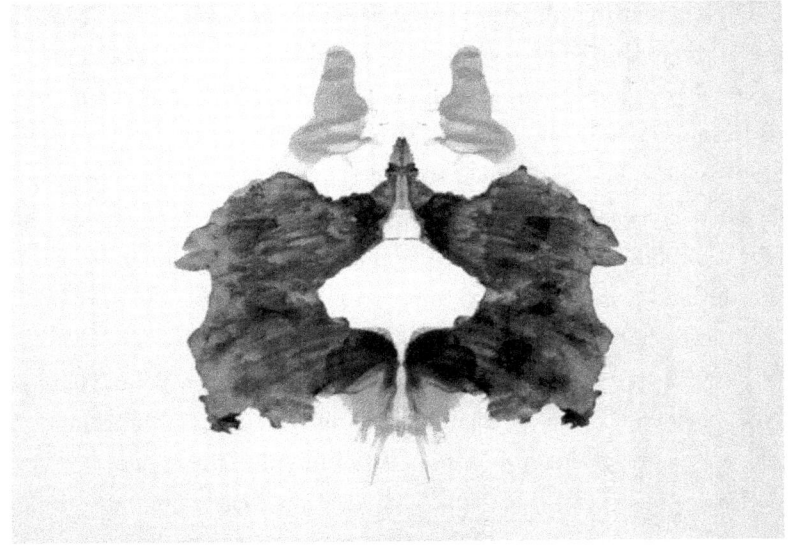

Singularity

81

When humans will merge with machines

THE GENERAL VIEW is that humans are still evolving and, according to some experts, probably quite rapidly. So where are we heading? Ray Kurzweil, an American inventor and futurist, has an interesting take on this question. He says that in the not-too-distant future, humans will merge with technology and biological evolution will become obsolete. This 'cutting-edge of evolution on our planet' will create something with smarter-than-human intelligence. He calls this new era 'the Singularity'.

In his book, *The Singularity is Near: When Humans Transcend Biology* (2005) Kurzweil says, 'As we get to the 2030s, the non-biological portion of our intelligence will predominate. By the 2040s it will be billions of times more capable than the biological part.'

Kurzweil divides the present and the future history of evolution into six epochs. 'The Singularity will begin with Epoch Five and will spread from Earth to the rest of the universe in Epoch Six,' he assures us. The six epochs are:

Epoch One: Physics and Chemistry. In this epoch, which started soon after the Big Bang, information was in atomic structure, especially in complicated, information-rich, three-dimensional structures of carbon atoms. The exquisite and delicate balance of the physical laws of the universe is precisely what is needed for the codification and evolution of information, resulting in increasing levels of order and complexity. Evolution, to Kurzweil, is a process of creating patterns of increasing order.

Epoch Two: Biology and DNA. In the second epoch, starting several billion years ago, DNA provided the precise digital mechanism for storing information and keeping it as a record of the evolutionary experiments of the epoch.

Epoch Three: Brains. The third epoch started when early animals developed the ability to recognise patterns, which still accounts for the vast majority of activities in our brains. Like earlier epochs, this epoch also provided a revolutionary or paradigm shift in the evolution of information: organisms now could detect information with their sensory organs and process and store that information in neural patterns in their brains.

Epoch Four: Technology. The pace of evolution of human technology is quickening. This quickening is exponential (when we plot growth against time on a graph, a J-shaped curve shows exponential growth; a straight line shows linear growth); for example, it took only fourteen years from the PC to the Web.

Epoch Five: The Merger of Human Technology and Human Intelligence. When the Singularity arrives, the human-machines will 'overcome age-old problems and vastly amplify human creativity'. Does this mean no nukes, no wars and the human-machine race lives forever in harmony and peace? Kurzweil is not so sure: 'But the Singularity will also amplify the ability to act on our destructive inclinations, so its full story has not yet been written.'

Epoch Six: The Universe Wakes Up. In the last epoch in 'the evolution of patterns of information', our civilisation will infuse the rest of the universe with its creativity and intelligence (we suppose if it doesn't destroy itself in Epoch Five). Of course, Kurzweil realises that this infusion can happen only when human-machines can circumvent the limit that the speed of light imposes on the transfer of information. When we are able to supersede this limit, 'the "dumb" matter and mechanisms of the universe will be transformed into exquisitely sublime forms of intelligence'.

'This is the destiny of the universe,' he declares.

82
Fish gods of the star Sirius

IN 1976 ROBERT Temple, an American author, published a book, *The Sirius Mystery: New Scientific Evidence of Alien Contact 5,000 Years Ago*, in which he claims that Earth has been visited by intelligent beings from the region of star Sirius. He says that the Siriusians were probably amphibians – 'a kind of cross between a man and a dolphin' – and left advanced astronomical knowledge that is still possessed by the Dogon people of Mali, West Africa.

The Sirius is a binary star system consisting of two stars orbiting each other. Situated in the constellation Canis Major (the Big Dog), it's only 8.6 light-years away. The larger of two stars is known as Sirius A (or Dog Star). It's the brightest star in the night sky (Venus and Jupiter are often brighter but they are not stars). In ancient Greece, the dawn rising of Sirius marked the hottest part of summer, which gave rise to the phrase 'dog days of summer'. Sirius A's faint companion was discovered in 1862. Known as Sirius B (nicknamed the Pup), it cannot be seen by the naked eye. Astronomers have found perturbations in the orbit of Sirius B and have wondered whether the system might have a third star, Sirius C. A recent search by the Hubble Space Telescope has not found any evidence to support the existence of a third star. It's highly unlikely that an Earth-like habitable planet orbits the youthful Sirius A; even if it did, it's most likely to develop primitive microbial life, not intelligent life. Sirius B is a white dwarf, an old dying star which is cooling down. White dwarfs have one of the densest forms of matter. No possibility of life there, either.

The French anthropologist Marcel Griaule and his student Germaine Dieterlen, who worked among the Dogon from 1931 to 1951, are known for their pioneering studies of the Dogon mythology. In a 1950 article they recorded that the men of the tribe held secret knowledge, which they revealed to Griaule as they accepted him as an honoured man of the tribe. The secret knowledge was about an invisible and super-dense companion of the star Sirius that moved around it every fifty years. A crude Dogon sand drawing shows the complete picture of the relative positions and movement of Sirius A, B and C. Sirius B was discovered in 1862, but Dogon legend is apparently centuries old. The Dogon legend also speaks about an 'ark' that descended to the ground amid a great wind. This legend is reflected in Dogon mask designs which resemble rudimentary rockets.

Sirius was the most important star to the ancient Egyptians; their calendar was based on the rising of Sirius. Armed with numerous anthropological, historical, literary and mythological sources, Temple tries to show that the Dogon were in fact the descendents of ancient Egyptian who fled during the tumultuous reign of pharaoh Akhenaten. He also presents a sophisticated patchwork of evidence to support his central theory.

Critics disagree with Temple's provocative theory that the Dogon preserve a tradition of what seems to have been an extraterrestrial contact. They say that Dogon legends were probably contaminated by their contact with Westerners. It's possible that information gained from Europeans was grafted on their existing legends. In 1978 the British astronomer and science writer Ian Ridpath wrote in the *Skeptical Inquirer* magazine: 'The whole Dogon legend of Sirius and its companions is riddled with ambiguities, contradictions and downright errors, at least if we try to interpret it literally.' In a review published in 1977 in *The International Journal of African Studies*, Ronald Davis of Western Michigan University noted that Temple's book 'is irritating because he evidently allows his enthusiasm to obscure his judgment and the rigor of his research ...*The Sirius Mystery* will be eagerly embraced by the lunatic fringe and ignored by serious students of West African culture and history.'

83

Unshrouding a mystery

THE SHROUD OF TURIN has been venerated for centuries as the cloth used to wrap the body of Jesus Christ after crucifixion. This 4.3-metre (14-foot) long piece of linen carries what appears to be the bloody imprint of a naked man lying with his hands crossed on his stomach. The image looks like a scorch and appears on only one side of the linen; it hasn't permeated the fibres. When first photographed in 1898 by Secondo Pia, an Italian lawyer and amateur photographer, the image resembled a photographic negative.

Because it first came to attention in 1357, the shroud has frequently been branded a 14th-century European forgery. In 1988 the Vatican agreed to a radiocarbon dating of the relic. Radiocarbon dating is a technique to measure the age of organic materials. An international team of scientists analysed a sample, about the size of a postage stamp, and concluded that it was created between 1260 and 1390. This led to the Archbishop of Turin, the custodian of the shroud, admitting that the shroud was a hoax.

But some myths are not easy to debunk. Many so-called shroud scholars have claimed that the sample used in the 1988 tests was cut from a medieval patch woven into the shroud to repair fire damage. The shroud was indeed damaged in a church blaze in 1532 and was restored by nuns by patching holes with a new material. In 2005 Raymond Rogers, a retired American chemist, claimed that a micro-chemical test performed on a piece of shroud the size of a grain showed that the shroud was between 1,300 and 3,000 years old.

The FOLLOWING PICTURE shows detail from Secondo's Pia's 1898 photograph of the Shroud of Turn.

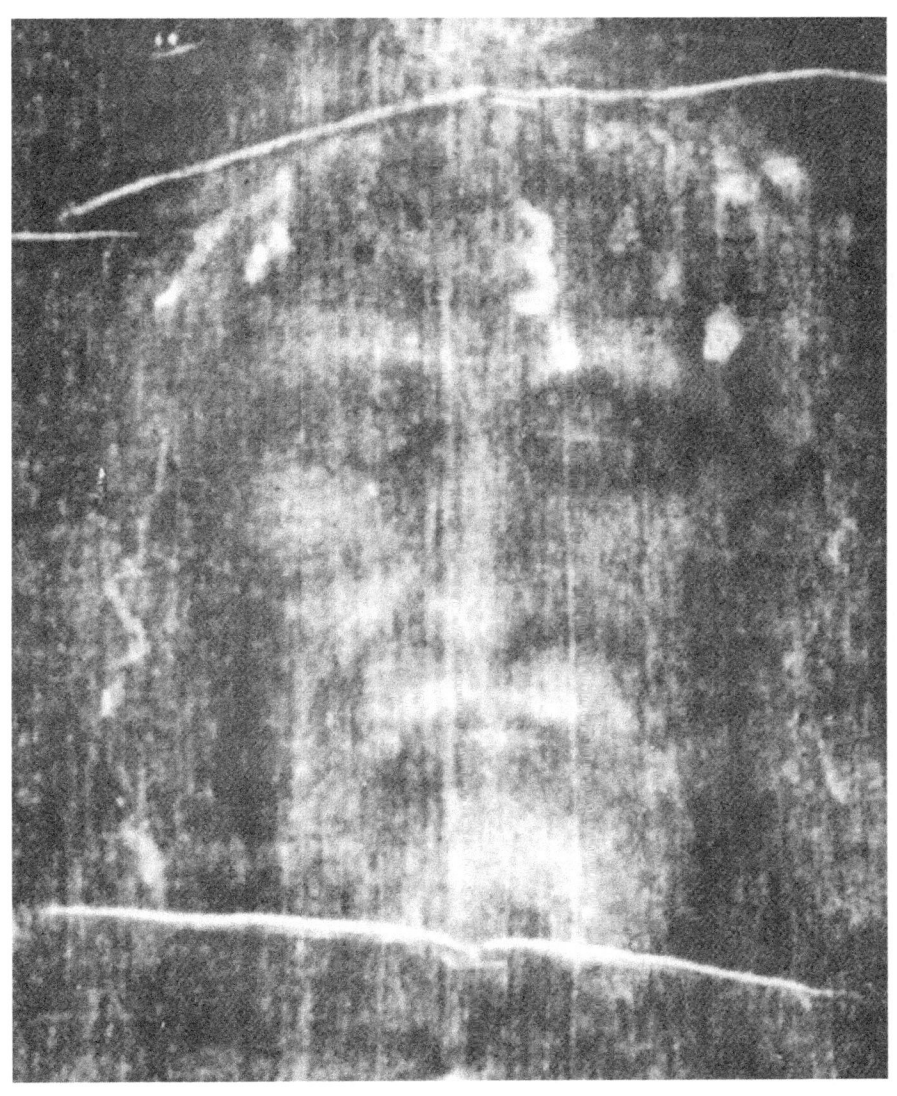

84
Can heavy drinking turn you into a 'crumbled black thing'?

KROOK, A RAG and bottle shop owner in Charles Dickens' novel Bleak House, was 'continually in liquor'. One day he mysteriously turned into a 'crumbled black thing'. Dickens describes what was left behind when Krook died without any apparent reason: 'Here is a small burnt patch of flooring; here is the tinder from a little bundle of burnt paper, but not so light as usual, seeming to be steeped in something; and here is – is it the cinder of a small charred and broken log of wood sprinkled with white ashes, or is it coal? O Horror, he IS here! and this, from which we run away, striking out the light and overturning one another into the street, is all that represents him.'

To kill heavy-drinking Krook, Dickens came up with the idea of spontaneous human combustion. The idea was not novel; it was known in Dickens' 19th-century world, but it fuelled the erroneous belief that heavy drinking could lead to spontaneous combustion. In spontaneous human combustion, the body is alleged to burst suddenly into flames without any outside source of heat and is reduced to a pile of greasy ashes in a few seconds. Paradoxically, other objects nearby remain relatively unburned; an example from another 19th-century novel, Jacob Faithful by Frederick Marryat: 'Nothing was burning – not even the curtain to my mother's bed appeared to be singed ... there appeared a black mass in the centre of the bed.'

In the early 1950s, Lester Adelson, a distinguished American pathologist, thoroughly investigated this macabre subject, and concluded that the cases of spontaneous human combustion could be readily explained as accidents or (rarely) as homicides, and dismissed the phenomenon as a 'relic of an age when love of the marvellous, the miraculous, and the seemingly inexplicable titillated the mind and imagination of scientists and layman ... It is not importable that the degree of burning was exaggerated by writers and sensation seekers to hep emphasise a point, a feature of news reporting not unknown today.'

Well, that feature of news reporting is still alive and kicking. Cases of spontaneous human combustion are even reported these days, especially when police and fire investigators have found burned corpses but no burned furniture. A few facts: the human body is mostly water and, except fats, tissues and methane gas, nothing within the body readily burns. The living human body has a temperature of about 38°C (100°F), but fire burns at more than 92°C (200°F). How a body could then suddenly self-ignite? Even if it does it could continue to burn only if the temperature is at least the ignition temperature or above.

The supporters of spontaneous human combustion give many reasons for this happening: excessive drinking of alcohol (inflammable gases are generated in the body after death), certain types of food (which produce a spontaneous chain reaction inside the digestive system), static electricity (which acts as a trigger of fire), nuclear energy (release within the body), psychic suicide (whatever that means) and the wrath of God. These theories are mere speculations as none of them has any scientific evidence to support it.

Sceptics say that almost all cases have simple plausible explanations such as the victim died of a heart attack and fell into an open fire or an electrical appliance. We humans are hard-wired to believe in the supernatural, and natural explanations don't titillate our minds.

85
It's impossible, indeed

TO DRAW A square equal in area to a given circle seems simple, but it has become a byword for impossibility. To square a circle is to do something that is considered to be impossible.

The problem was of extreme interest to people throughout the ages, not only to mathematicians but also the general public. Almost every great Greek mathematician tried in vain to solve it. The quest to find a solution continued into the eighteenth century. The Paris Academy received so many erroneous proofs that in 1775 it passed a resolution not to examine any proofs. The problem, at last, was 'solved' in 1882 when the German mathematician Ferdinand von Lindemann proved unequivocally that squaring the circle was an impossibility. His answer was based on the quirky nature of pi. All circles are similar and the ratio of the circumference to diameter is always the same number. This number is known as pi, one of the most important and ubiquitous numbers in mathematics.

Archimedes is now mostly remembered for the tale of his running naked through the street shouting, 'Eureka! Eureka.' But the greatest scientist, mathematician and mechanical genius of antiquity was the first to show that the problem of squaring the circle is equivalent to finding the area of a right-angled triangle whose sides are equal respectively to the circumference of the circle and the radius of the circle. Half the ratio of these two lines is equal to pi.

We know now that pi is an irrational number. It takes infinite digits to express it as a decimal number. It is impossible to find the exact value of pi; however, the value can be calculated to a very high degree of accuracy. Therefore, it's impossible to construct a square with the same area as a given circle using ruler and compasses alone.

86

The mystery of a supernatural star

When they heard the king, they departed; and behold, the star which they had seen in the East went before them till it came and stood over where the young Child was. When they saw the star, they rejoiced with exceedingly great joy. – Matthew 2:9-10

THE STAR OF Bethlehem leading the Magi to Jesus' birthplace is a standard symbol of Christmas. For two millenniums, astronomers, theologians, believers and sceptics have pondered the story of the star that is supposed to have announced the Christian era. Was there really a star – a holy light that guided the three wise men from the East to the manger in Bethlehem? Or was it a myth used as evidence by overzealous believers to confirm ancient prophecy? After all, no great king or conqueror of ancient history was born without some celestial spectacle being claimed to herald the event. Stars are said to have greeted the births of Alexander and Caesar.

St Augustine and other early Christian theologians accepted the star as one of God's miracles. 'When Copernicus, Kepler and Galileo ushered in the rise of empirical science, it became fashionable for Christian scholars to seek natural causes for events which the Bible clearly describes as supernatural,' says Martin Gardner, a well-known author of numerous books and a relentless fighter against pseudoscience.

One of the earliest and most obvious attempts to identify the star of Bethlehem was the suggestion that the Magi had seen the brilliant planet Venus, so often a splendid point of light in the morning or evening sky. But it is far too common an event to be a prophetic sign to the Magi.

Kepler was the first to identify the star with a precise event and date. In December 1603 Kepler was intrigued by the planetary conjunction of Jupiter and Saturn. With his characteristic patience and accuracy, he began computing the planetary positions at the time of the birth of Jesus. His calculations showed that there was a triple planetary conjunction of Jupiter and Saturn in 7 BC on 27 May, 5 October and 1 December. Critics of this theory say that there is a major flaw: the Bible refers specifically to a 'star', not a planet or a pair of planets. Besides, these conjunctions lasted only a few days, while the Magi's journey must have taken a few weeks.

Now, other theories also compete with Kepler's theory (some say that the great Kepler was only trotting out a theory that had been floating about since the 13th century). A recent suggestion is that the Magi were attracted to Bethlehem by a double eclipse of Jupiter and not by a star. The planet disappeared behind the Moon on 20 March 6 BC and then a month later on 17 April. The calculations show that the eclipse of 17 April occurred in daylight and would not have been visible as the guiding star to the Magi.

Or, was it a nova or supernova? A nova or supernova is an old star exploding and brightening near the end of its life. Both rise to full brilliance very quickly, often in a matter of few days, and then decline in brightness slowly. Novae are usually no brighter than ordinary brighter stars but supernovae sometimes become bright enough to be seen in the daytime. Most astronomers generally agree that the star of Bethlehem was not a supernova. Such stellar catastrophes are far too spectacular to not generally notice and, with the exception of Matthew, none of the historical writers of the time mention such a brilliant star near the time Jesus was born. However, according to Chinese and Korean astronomical chronicles, there is a record of the sighting of a bright nova in the Far East.

Many other less acceptable suggestions for the star have been offered. A spectacular meteor – a fireball or a meteor shower – could possibly agree with the behaviour of the star. Solar eclipses have also been suggested. There was a total solar eclipse of 20 January 2 BC and another on 1 January 1 BC. These events, though striking, would not be considered 'a star'. There is a record of a comet in 5 BC in ancient Chinese astronomical records, but it could not be the star of Bethlehem as in the Western tradition comets have always been considered harbingers of doom.

Some have even suggested that the star was not an astronomical event but a luminous phenomenon in the lower atmosphere such as *Aurora borealis* (a beautiful display of coloured lights in far northern skies). This theory argues that no celestial event could pinpoint an individual house as indicated in the Gospel of Matthew. Although a phenomenon like *Aurora borealis* might satisfy this requirement, it would hardly be called a star. But then in ancient times almost everything in the sky was called a 'star' of some sort: there were fixed stars, planets were wandering stars, comets were hairy stars or broom stars, meteors were falling stars and novae were new stars.

We may never learn the true identity of the star of Bethlehem. In a sense it hardly matters, for what is more significant about the star of Bethlehem is not whether it existed or where it was, but what it symbolises.

'Let the Bible be the Bible!' advises Gardner, 'It's not about science. It's not accurate history.'

Synchronicity

87
Coincidences: remarkable or random?

THE NAME OF a friend who you haven't seen for years suddenly pops up in your mind, you turn your computer on and there is an email from her. In your dream you see a lottery ticket and the next day you win a prize in a lottery. We all have had a premonition or a dream that later became true.

The Swiss psychologist Carl Jung coined the term 'synchronicity' to describe the simultaneous occurrence of events that were linked but couldn't be explained by the law of cause and effect. He said that such coincidences were meaningfully related; they were not random events happening in accordance without the laws of probability.

Jung was a colleague of Sigmund Freud but broke with him to work on his own psychological theory. In his theory, the psyche (the human mind and its deepest feelings and attitudes) has three layers: ego (which he identified with the conscious mind), the personal unconscious (personal reservoir of experiences unique to each individual), and the collective unconscious (a reservoir of our collective experiences as humans). The idea of collective unconscious makes Jung's theory unique. We can never be directly conscious of this spaceless, timeless realm of our experiences as a species, the theory says, but under certain states of altered consciousness, knowledge can slip through and allow us to glimpse a future event.

To find evidence for his idea of synchronicity, Jung analysed horoscopes of four hundred married couples. He could not find any significant statistical support for the existence of a connection without any cause between horoscopes and actual marriages, yet he clung to the idea. Jung had a lifelong interest in the paranormal and his idea of meaningful coincidences made some sense in his nineteenth-century world, a world saturated with spiritualism and the occult.

No scientific evidence has ever been found for synchronicity. 'Meaningful coincidences' are not remarkable but random coincidences. Nevertheless, the New Age spirituality seems to be synchronous with synchronicity because it alludes to a vague spiritual connection between mind and matter.

Tachyons

88
Faster than fact

TACHYONS ARE THEORETICAL particles that travel faster than light.

If Einstein had read the following relativistic limerick, he would have said that Miss Bright had 'no possibility of existence', the phrase he used for speeds greater than the speed of light in his famous 1905 paper on special relativity.

There was a young girl named Miss Bright
Who could travel much faster than light.
She departed one day,
In an Einsteinian way,
And came back on the previous night.

In 1962 a group of physicists went against Einstein's speed rule and proposed particles that could travel faster than light. They said that Einstein's rule didn't apply to these particles. In 1967 the American physicist Gerard Feinberg named them tachyons (after Greek *tachus*, 'swift').

Weird tachyons, theoretically speaking, are always travelling faster than light — they're 'born' with speeds greater than light. When tachyons lose energy, they gain speed. When they gain energy, they slow down. Infinite energy is required to slow down a tachyon to the speed of light.

Most physicists have given up the idea that tachyons might be real. Nevertheless, they keep on popping up in many new physics theories. For the time being Einstein's speed rule remains sacrosanct and tachyons don't exist. But it hasn't stopped them entering into the New Age world. Beware of enterprising New Agers who claim that they've harnessed the power of tachyons.

Teleportation

89

Beam me up, Scotty!

MANY OF US were introduced to the idea of teleportation by the popular *Star Trek* television series in the late 1960s. We watched in amazement as Captain Kirk of the fictional *Starship Enterprise* dematerialised in one place and instantly reappeared on another planet. When he wanted to be transported back to the spaceship Captain Kirk would command his chief engineer Montgomery Scott, 'Beam me up, Scotty!'

The term 'teleportation', however, originated with Charles Fort, a 19th-century American writer who was obsessed with curious phenomena which science could not explain. He believed that objects could be instantaneously moved between Earth, which he believed was nearly stationery, and other heavenly bodies by teleportation. The term 'Fortean' now means 'relating to paranormal phenomena'.

According to the laws of science, to teleport an object we need to know the exact location of every atom in the object. This idea violates the uncertainty principle which says that we cannot know both the precise location and the velocity of an electron. Therefore, according to classical physics, teleportation of large objects or humans is impossible. But quantum mechanics has different rules, which apply only to processes that occur at the level of individual atoms. One of the bizarre features of quantum mechanics is quantum entanglement. All elementary particles such as electrons and photons vibrate. Consider the case of two electrons. Placed together they vibrate in unison. Place them apart, as far as another galaxy, and if you vibrate one of them, the other will immediately know the nature of its partner's vibrations and dance to the same tune. Somehow the information between the two electrons is being transferred. Einstein called it 'spooky action at a distance' because the transfer of information could only be explained by assuming that it was travelling faster than light.

Could quantum entanglement be used to transfer information faster than light? The laws of quantum mechanics prohibit it, but it could be used to transfer information from one particle to another particle at a speed slower than that of light. In quantum teleportation, only the quantum state is teleported, not exact particles. So, quantum teleportation is not really instantaneously sending one particle, say an electron, from one place A to another place B. But the quantum states of the electron at A and B are indistinguishable. It's not really 'quantum faxing'; in faxing it's easy to tell the difference between the original and the copy. In quantum teleportation, there is no difference between the original and the copy.

In 1999 an international team of physicists demonstrated that quantum teleportation is indeed practically possible, but only if the original is destroyed. The team has been successful in teleporting photons, the particles of light. The idea of teleporting a molecule remains a fantasy, but scientists have not ruled it out in the far future.

The most important application of quantum teleportation is in the field of quantum computing. Ordinary computing is based on the notion of bits. An ordinary bit can store only one number (0 or 1) at a time. In quantum state a particle can occupy two states at the same time, so a quantum bit, or qubit, can store two numbers at the same time. Thus each qubit doubles the size of a computer. One day quantum teleportation and quantum computing would make the transfer of information incredibly fast. Beam my file up, Scotty!

90
Untapped resources of the brain

YOU HAVE PROBABLY heard that we use only 10 per cent of our brains? There is absolutely no scientific evidence, even of moderate quality, to support this absurd claim.

In recent years, neuroscientists have scanned the brain with sophisticated big-name machines such as electroencephalography (EEG), magnetoencephalography (MEG), computerized axial tomography (CAT) and positron emission tomography (PET) and functional magnetic resonance imaging (fMRI) and have pinpointed numerous psychological functions to its specific parts. Their scans have not revealed any portions of the brain which are in a vegetative state. Besides, if 90 per cent of our brains was really doing nothing, there would be large areas of dead cells in our brains. No autopsy has ever revealed it to be true.

At any given time, not all neurons, the basic working units of the brain, are active; but no neuroscientist has ever found that 9 per cent of our brain is perpetually on vacation. Even at rest, the brain works at full capacity. Brain scans show that our brains have a 'default network', a sophisticated network of brain areas that remains active when we are supposedly doing nothing. Of course, some parts of the brain are more active than others at any given time or during a particular activity.

For our body, the brain is an expensive organ to maintain; it uses too many resources: about 20 per cent of our body's daily calories intake. Evolution (or if you prefer intelligent design) would not have allowed such a wasteful organ to survive.

Yet, the myth of 10 per cent brain refuses to die.

A survey conducted by Paul Howard-Jones of the University of Bristol reveals that 59 per cent Chinese teachers agree that we use only 10 per cent of our brains. A study by Sanne Dekker of VU University Amsterdam informs that this figure is 48 per cent and 46 per cent for teachers in the UK and the Netherlands respectively.

Not convinced by these results? Conduct your own mini survey. Ask 10 people in your workplace or anywhere else and you would be surprised by the high percentage of people who believe in this myth. When someone tells you that we use only 10 per cent of our brains, they are probably using only 10 per cent of their brains.

Where does this myth come from? Some suggest that it came from William James, often referred to as the father of American psychology, who in 1907 wrote in an essay, titled 'Powers of Men': 'As a rule men habitually use only a small part of the powers which they actually possess and which may they use under appropriate conditions ... We are making use of only a small part of our possible mental and physical resources.' In 1936, in his preface to Dale Carnegie's How to Win Friends and Influence People, one of the best-selling self-help books all time, the famous journalist Lowell Thomas attributed the claim 'we use only 10 per cent of the brain' to William James.

The myth has been a boon to self-help gurus. Probably they would have to invent it if it didn't exist. A whole industry is based on this myth.

To be fair, the myth has an uplifting advantage. 'The 10 per cent myth has undoubtedly motivated many people to strive for greater creativity and productivity in their lives – hardly a bad thing,' observes Barry L. Beyerstein, an American psychologist. 'The comfort, encouragement and hope that it has engendered helps to explain its longevity.' There is another reason for its persistence: blockbuster movies like Lucy, released in 2014, are helping to perpetuate it.

Tesla's death rays

91
Powerful enough to destroy 10,000 aeroplanes

THE SERBIAN-AMERICAN scientist Nikola Tesla was a genius so ahead of his time that his contemporaries failed to understand his groundbreaking inventions. An inventor of dazzling brilliance, he invented and developed AC power, induction motors, dynamos, transformers, condensers, bladeless turbines, mechanical rev counters, automobile speedometers, gas-discharge lamps that were the forerunners of fluorescent lights, radio broadcasting and hundreds of other things (the number of patents in his name exceeds 700). However, these 'practical inventions' were limited to the short period from 1886 to 1903.

In later years of his life, Tesla was a favourite of newspaper reporters who revelled in recounting his incredible inventions. On his 78th birthday, Tesla told a New York Times reporter that he had invented a death ray powerful enough to annihilate an army of 10,000 planes and 1 million soldiers instantaneously. The next day, July 11 1934, the paper ran a story that was headlined in the style of the time:

TESLA, AT 78, BARES NEW 'DEATH-BEAM'
INVENTION POWERFUL ENOUGH TO DESTROY
10,000 PLANES 250 MILES AWAY, HE ASSERTS.
DEFENSIVE WEAPON ONLY
SCIENTIST, IN INTERVIEW, TELLS OF APPARATUS
THAT HE SAYS WILL KILL WITHOUT TRACE.

The story referred to Tesla as 'the father of modern methods of generation and distribution of electrical energy', and quoted him as saying that this latest invention of his would make war impossible: 'It will be invisible and will leave no marks behind it beyond evidence of destruction. This death-beam would surround each country like an invisible Chinese Wall, only a million times more impenetrable. It would make every nation impregnable against attack by airplanes or by large invading armies.'

On his 84th birthday, Tesla declared that he stood ready to divulge to the United States government the secret of his 'teleforce', with which aeroplane motors would be melted at a distance of 400 kilometres, so that an invisible wall of defence would be built around the country. He said that this teleforce was based on an entirely new principle of physics that 'no one has ever dreamed about', and would operate through a beam one-hundred-millionth of a square centimetre in diameter. The voltage required to produce this beam would be about 50 million volts, and this enormous voltage would catapult microscopic electrical particles of matter on their mission of defensive destruction, he added.

Tesla probably conceived the idea for his death ray at Wardenclyffe, Long Island, New York, where in 1902 he built a 57-metre 187-feet) tower and laboratories to experiment on radio waves and on transmitting electrical power without wires. The tower's steel shaft ran 36 metres underground, and it was topped with a 55-tonne, 20-metre diameter metal dome. This experimental facility had the financial backing of the legendary investor J. Pierpont Morgan. However, Morgan pulled out of the venture even before construction was complete. The tower was abandoned in 1911 and demolished in 1917.

The popular story that Tesla tested his death ray one night in 1908 goes something like this. In 1908, Arctic explorer Robert Peary was making the second attempt to reach the North Pole, and Tesla requested him to look out for unusual activity. On the evening of 30 June, accompanied by his associate George Scherff atop the Wardenclyffe tower, Tesla aimed his death ray towards the Arctic, to a spot west of the Peary expedition. Tesla then scanned the newspapers and sent telegrams to Peary to confirm the effects of his death ray, but heard of nothing unusual in the Arctic. When Tesla heard of the Tunguska explosion (*p. xxx*), he was thankful no one was killed, and dismantled his death ray machine, feeling it was too dangerous to keep.

In a letter to *The New York Times* on 21 April 1907, Tesla wrote: 'When I spoke of future warfare I meant that it would be conducted by direct application of electrical waves without the use of aerial engines or other implementation of destruction … This is not a dream. Even now wireless power plants could be constructed by which any region of the globe might be rendered uninhabitable without subjecting the population of other parts to serious danger or inconvenience.' Though he believed that it was 'perfectly practicable to transmit electrical energy without wires and produce destructive effects at a distance', there is no evidence that Tesla used the Wardenclyffe tower for his experiments on the death ray.

The death ray may have been a plausible dream, but it was not a reality. Tesla never got the opportunity to test his plans. The Tunguska story seems improbable for another reason. Tesla could not have heard about the Tunguska event before 1928, when stories about it appeared in the American newspapers. Also, there is no record of Tesla's request in Peary's accounts of his expedition. The story has simply been conjured up by joining the dots – Tunguska, Tesla, Peary – with 1908. But the dots do not interrelate.

Time reversal

92

Can time go backwards?

HERACLITUS, A BRILLIANT and highly original Greek thinker who flourished in the fifth century BC, believed in universal change: permanence is an illusion and everything in the world is constantly in change. 'You can't step into the same river twice, for fresh waters are ever flowing in upon you,' he said. The image of time as a river is arguably the oldest metaphor of time, and Heraclitus' river of time could only flow in one direction: towards the future. Like an arrow time flies in one direction only. Why?

First, what's time, in a scientific sense? For physicists, time is simply what clocks measure; for example, the speed of light (300,000 kilometres or 186,000 miles per second), or how frequently our planet spins (one rotation per day). Mathematically, time is one-dimensional space. It may be treated as another dimension of space, it's not like space. 'In a very precise sense, time is the direction within spacetime in which good prediction is possible – the direction in which we can tell the most informative stories,' says Craig Callender, a professor of philosopher at the University of California. 'The narrative of the universe doesn't fold in space. It unfolds in time.'

But why does the narrative fold in one direction only? The answer lies in entropy, the measure of the disorder of a system.

Rudolf Clausius, a 19th-century German theoretical physicist, was one of the founders of thermodynamics, the study of work, heat and other forms of energy. He developed thermodynamics from two laws. The first law of thermodynamics is the law of conservation of energy: energy is neither created nor destroyed. The second law, which he formulated himself, is the law of dissipation of energy: heat does not flow spontaneously from a colder to a hotter body.

Like Newton's laws, the second law is a scientific attempt to explain the universe. The law gives a direction to time – basically, it says that many natural processes are irreversible. One consequence of this irreversibility is the 'arrow of time'. Now you know why scrambled eggs can't be unscrambled.

In 1865 Clausius introduced the term entropy as a measure of disorder or randomness of a system. The more random or disordered a system is, the greater the entropy. In a closed system, entropy never decreases and must ultimately reach a maximum. Entropy is continually increasing in the universe. But because the universe is a closed system, once all the energy in the universe is converted into heat, there will be no energy available for work. This would bring 'the heat death of the universe'.

It follows from the idea of 'the heat death of the universe' that that the universe cannot have been drawing on this finite supply of energy for all eternity. Or, the universe has not been around forever. The universe – that's space and time – began with the Big Bang about 13.7 billion years ago. The Big Bang is the moment where scientists' understanding of time ends. No one knows what happened before the Big Bang.

Anyway, time marches forward because entropy in the past was lower than now. This idea also explains why we remember the past but not the future. Because of low entropy the past was orderly. Similarly, because of high entropy the future would be disorderly. Reliable memory requires order, not disorder or random fluctuations that would in no way be related to the past.

Time travel

93

Travelling back and forth in time

TRAVELLING THROUGH THE three dimensions of space is easy; we can travel backwards, forwards and sideways. But when it comes to travelling in the fourth dimension we are stuck, we can only move forward. Why? Einstein has the answer.

The theory of general relativity brought space and time in a single four-dimensional arrangement called spacetime. It's easy to understand spacetime if we imagine it as a rubber sheet. This sheet is stretched out flat if there are no objects on it, but an object will warp this sheet. We say that gravity warps spacetime.

In the theory of special relativity Einstein said that time is not an absolute quantity. Our measurements of time are affected by our motion. The rate at which clocks run depends upon their relative motion. A person running away from a clock would observe it to move more slowly than his or her own clock. The theory also says that the mass of a moving object increases as its speed increases. At the speed of light, which is about 300,000 kilometres per second (186,000 miles per second), the mass becomes infinite and therefore nothing can move faster than light.

Theoretically, a spaceship travelling nearly at the speed of light would take nine years by Earth's calendar to make a return trip to Centauri, the next nearest star after the sun; but because of relativistic time changes, on return to Earth the crew would find that many decades have gone by. However, the crew would notice no change on the spaceship. From their point of view, the spaceship is stationary and Earth is moving at almost the speed of light and time on Earth slows down.

Relative time poses an interesting paradox: if one twin goes on a high-speed space journey, she would return younger than her sister who stayed home. It defies common sense, but Einstein once remarked that it was the common sense which once objected to the idea that Earth is round (*story 19*).

As long as Einstein's theory is supreme, time travel into the past or future would remain in the realm of science fiction. 'If time travel is possible, then where are the tourists from the future?' wonders the celebrated physicist Stephen Hawking known to everyone for his book, *A Brief History of Time*. Probably, their spaceships have been stuck in wormholes (*story 98*), the 'science fiction' of theoretical physicists who love to imagine 'what if' scenarios to understand the true nature of the universe.

Trepanation
94
You need it like a hole in the head

THE BEATLES' PAUL McCartney, in an interview in *Musician* magazine (October 1986), recalled John Lennon once asking him, 'You fancy getting the trepanning done?' McCartney asked, 'Well, what is it?' Lennon replied, 'You kind of have a hole bored in your skull and it relieves pressure.' The story is not surprising given that in the psychedelic 1960s millions of young people were experimenting with mind-altering experiences.

The idea of deliberately drilling a hole roughly the size of a man's watch into the skull to heighten consciousness seems barbaric, but archaeologists say that trepanation (from Greek *trupanon*, 'to bore') is the oldest surgical practice. A trepanned skull found in France was estimated to be 7,000 years old. In the fifth century BC, the Greek physician Hippocrates, regarded as the father of Western medicine, wrote detailed instructions on how to perform trepanation to relieve pressure on the brain caused by disease or trauma. In some cultures, it was practised to release evil spirits.

This mix of magic and medicine has always fascinated people and the practice has not died out completely. Amanda Fielding, a British painter, attracted the attention of the world's media when in 1970 the 27-year-old filmed herself performing self-trepanation. The film shows her standing in front of a mirror wearing a white frock. She takes a dentist's electric drill and starts drilling through the front of her shaved forehead.

Fielding (now Lady Neidpath, she runs the Beckley Foundation which does research on consciousness and its changing states) says that having a hole in her head allows more blood to reach her brain, which increases brain metabolism and expands her consciousness.

'Hole in the head' is not a third-eye to total perception. It's a fallacy. Trepanation is a dangerous operation — there are risks of blood clots, brain injuries and infections leading to meningitis or death – and no self-respecting surgeon would ever dream of performing it. There are no known psychological benefits associated with this crude ancient procedure.

Tunguska Event

95

The riddle of a fireball

ABOUT 7.14 A.M., 30 June 1908. The Central Siberian Plateau near the Stony Tunguska River, a remote and empty wilderness of swamps, bogs and hilly pine and cedar forests. Not a soul in sight for scores of kilometres. The eerie silence is punctuated by the shuffle of the hoofs of reindeer grazing in the morning sun.

Suddenly a blindingly bright pillar of fire, the size of a tall office building, races across the clear blue sky. The dazzling fireball moves within a few seconds from the south-southeast to the north-northwest, leaving a thick trail of light hundreds of kilometres long. It descends slowly for a few minutes and then explodes above the ground. The explosion lasts only a few seconds, but it is so powerful that it can be compared only with an atomic bomb – 1,000 Hiroshima atomic bombs. A dark mushroom cloud almost reaches to space.

The explosion flattens an area of remote Tunguska forest bigger than Greater London, stripping millions of ancient trees of leaves and branches, leaving them bare like telegraph poles and scattering them like matchsticks. A black rain of debris and dirt follows.

Hundreds of miles away, the Trans-Siberian Express rattles wildly on its newly built tracks. Tremors are registered in distant St Petersburg. For many days after the explosion unusually bright night skies are seen in many parts of Europe.

In 1908 Russia was a country caught in political unrest and social upheaval. Besides, the impact site was in Russia's most remote and rugged region and no effort was made to launch a scientific expedition there.

The first scientist to visit Tunguska was Leonid Kulik, a meteorite expert. When he visited the site in 1927 he saw an oval plateau 70 kilometres (43 miles) wide where the forest had been flattened, all the trees stripped of their branches and scattered like matchsticks pointing away from the direction of the blast. He believed that the devastation had been caused by a meteorite. He visited the impact site again three times but could not find any crater.

If it were not a giant meteorite, then what caused the great Siberian explosion? The controversy about the Tunguska fireball continues to this day, and there is no shortage of attempts to explain the cataclysmic explosion.

A computer used to simulate the Tunguska explosion has pointed an accusing finger at a stony asteroid that exploded 8 kilometres (5 miles) above the ground, the same height at which the Tunguska object is believed to have exploded. What about the rubble from the exploding asteroid? The blast was so strong that it turned the pieces of asteroid into vapour. Only the microscopic gravel would fall into the swamps below. It was nineteen years after the explosion that the site was actually searched and by then the gravel would disappear in rain like salt in water.

A comet head that exploded in the air before hitting Earth might have caused the Tunguska explosion. As a comet head is a 'dirty snowball' made of ice and dust, on explosion it would create a fireball and blast wave but no crater. But sceptics of the comet theory doubt that a body large enough to cause the Tunguska explosion would have remained unnoticed by astronomers.

The asteroid and comet theories appear to be the most promising but, for many, an asteroid or a comet does not really answer the riddle of the Tunguska explosion, the biggest impact in recorded history.

Another interesting theory is that the explosion was caused by a mini black hole that entered Earth and passed straight through it in about fifteen minutes and exited through the North Atlantic. Though the existence of massive black holes has been proved, mini black holes are still a figment of scientists' imagination.

A bizarre idea is that the explosion was caused by a 'cosmic visitor' – an extraterrestrial spacecraft, cylindrical in shape and propelled by nuclear fuel. Because of a malfunction the spacecraft hit Earth and within a fraction of a second, the spacecraft and its occupants were vaporised in a blinding flash of light. The ETs had come to collect water from Lake Baikal, which is the largest volume of surface freshwater. Apparently, the ETs were from a parched planet and very, very thirsty.

UFO fans prefer the theory that a faulty flying saucer was deliberately crashed into the sparsely populated Siberia by its considerate extraterrestrial crew to save human lives. Some add another twist: the explosion was caused when the engine of a spaceship blew up. The spaceship had left two thousand years earlier and was returning home, but missed the runaway.

And, of course, there is a conspiracy theory as well: it was an early experiment in nuclear physics which got out of hand. The experiment on 'death rays' was conducted by Nicola Tesla, the enigmatic American scientist who invented and developed AC power and numerous other things (*story 91*).

96
Do they exist? Try the Santa Claus hypothesis

THE TERM 'UFO' (unidentified flying object) was suggested in the mid-1950s by the United States Air Force. The term 'flying saucer' was not considered accurate, since many sightings had very natural explanations, while others did not.

UFOs are so infrequent that they are unique to most observers. In that sense they are truly encounters of the UFO kind, but they are not encounters with extraterrestrials. 'U' in UFO simply means 'unidentified'; it doesn't suggest 'extraterrestrial'.

Most UFO fans believe that extraterrestrial intelligent beings are visiting Earth. They also believe that governments are covering up this fact because they know it would trigger a panic; governments are afraid of admitting something that is beyond their control.

Most of these UFO fans have also claimed to have seen a UFO. What have they seen? Planes, jets, helicopters, balloons, strange flocks of birds. Unusual light patterns caused by astronomical and meteorological phenomena. Optical illusions caused by smoke and dust. Psychological delusions. Deliberate hoaxes.

Many people who have had vivid memories of close encounters with UFOs are not convinced by these explanations. Scientists say that some UFO events are attributable to physical, electrical and magnetic phenomena in the atmosphere. These events create regions of electrically charged plasma which appear as bright, fast-moving objects to observers. They are probably caused by a meteorite entering the atmosphere, neither burning up completely nor impacting, but forming buoyant plasma. Sometimes the field between certain charged buoyant objects forms an area, often triangular, which does not reflect light. This explains why some UFOs are described as black spaceships, often triangular, and up to hundreds of metres in length. The events cannot be detected by radar.

The close proximity of plasma related fields can adversely affect a vehicle or person. It has been medically proven that local electromagnetic fields can cause responses in the temporal lobes of the human brain. These result in the observer sustaining (and later describing and retaining) his or her own vivid, but mainly incorrect, description of what is experienced.

Carl Sagan of *Cosmos* (the celebrated book and TV series) fame gives an interesting explanation: 'If each of a million advanced technical civilisations in our galaxy launched at random an interstellar spacecraft each year, our solar system would, on average, be visited once every 100,000 years.'

He employs his Santa Claus hypothesis to explain UFOs. The hypothesis maintains that in a period of eight hours or so on 24–25 December of each year, an elf visits 100 million homes in the United States. If this elf spends one second per house, he has to spend three years just filling the stockings in all the houses. Even with relativistic reindeer, the time spent in 100 million houses is three years and not eight hours. 'We would then suggest that the hypothesis is untenable,' he says. 'We can make a similar examination, but with greater certainty, of the extraterrestrial hypothesis that holds that a wide range of UFOs viewed on the planet Earth are space vehicles from planets of other stars.'

The stories of UFOs began with the now-famous Roswell incident. On 8 July 1947, the *Roswell Daily Record* broke the news of a cosmic encounter in New Mexico. The story, headlined 'RAAF Captures Flying Saucer on Ranch in Roswell Region', was based on a press release issued by Roswell Army Air Field (RAAF). When sheep rancher Mac Brazel was making rounds at a ranch near Roswell, he found some wreckage consisting largely of rubber strips, wood sticks, tinfoil, plastic, tape with some strange markings that resembled 'hieroglyphics', and very tough paper. Brazel was struck by the unusual nature of the debris. After a few days he drove into Roswell, where he reported the incident to the Sheriff, who reported it to the RAAF. The RAAF immediately issued a press release stating that the wreckage of a flying saucer had been recovered. The news caused a sensation around the world, but it was short-lived. Within hours the RAAF announced that remains of a weather balloon had been mistakenly identified as the wreckage of a flying saucer.

Another story that was not published in the paper, but that some town folks knew about, came from witnesses who had seen the wreckage. They claimed that they had seen alien 'bodies' nearby, describing them as a little more than a metre (3 feet) tall, with bluish skin colouration and no ears, hair, eyebrows or eyelashes. The RAAF explained these 'aliens' as dummies dropped from high altitude balloons to study the results of the impact. And that was the end of the excitement.

No one ever talked about this episode, at least until the publication in 1980 of a book, *The Roswell Incident*, which came to the dramatic conclusion that the United States government had found and removed the remnants of the UFO crew—several little alien bodies. This book was the genesis of a UFO myth and a conspiracy theory that the government had conspired to cover up the fact that an alien ship had landed at Roswell.

The truth is far less exotic: what actually happened was that people who saw the dummies mistook them for aliens.

97
Cold comfort

LINUS PAULING IS the only person to have been awarded two unshared Nobel Prizes – the 1954 Nobel Prize in Chemistry and the 1962 Nobel Peace Prize. Yet he is now mostly known for popularising the controversial contention that large doses of vitamin C are effective in the prevention of the common cold. In 1970 Pauling, who took 12,000 milligrams of vitamin C daily, published a book, Vitamin C and the Common Cold, in which he encouraged people to take 1,000 milligrams of the vitamin daily (the current recommended daily allowance is only 60 milligrams; a large glass of orange juice contains about 100 milligrams). This highly popular and influential book's recommendation was not based on much scientific evidence.

Common cold is not a precisely defined disease, yet most of us are familiar with its symptoms and believe that vitamin C can kill the cold. A 2007 Cochrane Review has shown that large doses of vitamin C have little preventive effect and are not effective in dealing with the symptoms. Based on best available evidence, Cochrane Reviews are the most respected and impartial evaluations of medical research.

The 2007 review analysed 30 clinical trials carried out around the world over six decades and involving more than 11,000 patients who took 200 milligrams or more of vitamin C daily. It revealed that those who were taking daily dose of the vitamin were just 2 per cent less likely to catch cold, which would equate to the average person suffering a cold 11 days a year instead of 12. 'It just doesn't make sense to take vitamin C 365 days a year to lessen the chance of catching the cold,' the review advised. However, vitamin C supplements may lower the risk of catching cold among marathon runners, skiers and others exposed to extreme cold or stress for short periods. Some medical experts suggest that vitamin C is of little help even in such extreme situations; it is likely to benefit only those lacking it in the first place.

98
A tunnel from here to eternity

IMAGINE A SPHERICAL opening in a park, say in Sydney, which you could enter and walk a few steps through to New York City's Central Park. Travel for just a few hours through the tunnel, you could even emerge in another part of the universe many light years away. More exotic than Alice's rabbit hole, wormholes, as tunnels in spacetime are called, are no longer considered an 'unscientific' proposition.

American astrophysicist J. Richard Gott is an expert on black holes, regions of space where the gravity is so strong that nothing, even light, can escape. While thinking about wormholes, he says, has no practical value, it helps scientists to explore the fundamental questions of cosmology: how the universe began, how it works and how it might end.

The idea of wormholes comes directly from Einstein's theory of general relativity which says that extremely dense objects stretch the fabric of space and time around them. If the density of an object approaches infinity, such as a black hole, that stretch can become a tear. This tear or wormhole could act as a tunnel to other parts of the universe making virtually instantaneous travel possible.

The laws of physics allow wormholes but it doesn't mean that one day you would be able to travel through them. To build a wormhole would require an unusual form of energy called negative energy. Again, the laws of physics allow negative energy but they also limit its behaviour. Negative energy exists but only in minute quantities.

Scientists estimate that to create a 1-metre wormhole would require fabulous amounts of negative energy, which would require matter equivalent to the mass of Jupiter to produce.

This 'Jupiter' would have to be made of negative matter to produce negative energy. Some future advanced civilisation might have the kind of technology to create a wormhole, but until then we will have to borrow Dr Who's time machine, Tardis (time and relative dimensions in space), if we want to travel to other parts of our universe or other parallel universes (*story 66*).

'Wow' signal

99
A missed phone call from ET?

IT CAME FROM outer space and lasted only 72 seconds. On the computer printout, it simply appeared as 6EQUJ5 – a code that revealed it was a strong, intermittent radio signal confined to a narrow band of frequency. It was so unusual that it caused an excited astronomer to scrawl 'Wow!' in the margin of the printout, a label that is inextricably linked to it.

What was it? A message from an alien intelligence? A momentary hiccup from a cosmic event, or a polluting burp from a terrestrial transmission? Decades on, no one knows what really created the signal, and the debate continues.

The Big Ear radio telescope at the then Ohio State University Radio Observatory had been involved in the search for extraterrestrial intelligence since 1973. On the night of 15 August 1977 its 79-metre dish was tuned to 1,420 megahertz, the frequency of hydrogen atoms. Hydrogen is the most common element in the universe; many people believe that the aliens might choose this frequency to broadcast their presence to us.

The Big Ear used a computer to record the signals. The computer printouts had to be examined manually by volunteers as the observatory did not have any funding for the program.

Jerry Ehman, a professor at Ohio State University, took the responsibility for this task. A few days after

15 August, he began his routine review of the printouts, not expecting to find anything unusual. As he worked his way through the reams of paper for the night of 15 August, he was astonished to see the string of numbers and characters 6EQUJ5 on the printout. It represented a burst of radio waves, like a thunderclap in the middle of a piece of quiet music. Ehman immediately recognised it as a narrow-band signal confined to a frequency around 1,420 megahertz. Without thinking, he wrote 'Wow!' and circled the string 6EQUJ5 on the printout. 'It was the most significant thing we had seen,' recalls Ehman.

For a month following the discovery, he and his colleagues looked for the signal again at least fifty times, but found nothing.

Each character of the alphanumeric code used by the observatory's computer represented the strength of a received signal over twelve seconds. The code 6EQUJ5 revealed that the 'Wow' signal rose and fell over the course of 72 seconds.

The fact that the signal rose and fell over the course of 72 seconds is intriguing. The Big Ear was a fixed telescope (it was demolished in 1998), and Earth's daily spin allowed it to pick up cosmic signals from a tiny angular section of the heavens. The string 6EQUJ5 shows that as the radio source passed by, its intensity rose as Earth's spin brought it within the telescope's range, reached a peak in the centre, and then faded away. For the Big Ear, this rise and fall should last 72 seconds, and that's what happened. If the signal were from a terrestrial source, it would suddenly flood the telescope and then switch off after some time.

A terrestrial signal also scores low probability for another reason: radio transmissions in the frequency band around 1,420 megahertz are prohibited by international agreement. Ehman and his colleagues also looked at other possible sources of the signal such as planets, asteroids, stars, satellites, aircraft and spacecraft but ruled them all out.

So the evidence stacks up in favour of an alien source. Not quite. The Big Ear used two funnel-shaped metal structures called horns, situated side by side, to collect radio waves focused by the dish. The path of the radio waves collected by a horn is called a beam. As Earth rotates, any cosmic radio source would be seen first in beam one (for 72 seconds) and then – about three minutes later – in beam two (also 72 seconds). Ehman's printout showed it only on one beam, instead of the two beams expected. The computer was not programmed to identify whether the signal came from the first or the second beam. It sounds suspicious.

Perhaps the aliens turned off their transmitter after three minutes and went on holiday. Was the signal a wish-you-were-here postcard? No one knows. Unlike alien astronomers, terrestrial astronomers are workaholics.

From observatories around the world they have performed more than hundred searches of the same region of the sky. No report yet of an astronomer running naked through the street, shouting 'Wow! Wow!'

Ehman also believes that even if the Wow signal were an alien signal, they would do it far more than once: 'We should have seen it again when we looked for it fifty times.' Other astronomers agree, but the ET fans don't.

THE FOLLOWING PICTURE shows a section of the computer printout of Jerry Ehman's 'Wow!' and the circling of 6EQUJ5 (Image: Jerry Ehman, North American Astrophysical Observatory, the continuing organisation of the Big Ear Radio Observatory)

100
Not for harnessing, either by scientists or 'energy healers'

THE CONCEPT OF zero-point energy belongs to quantum mechanics textbooks, not to the handbooks of practitioners of alternative medicine or proponents of perpetual-motion machines and other free-energy devices.

First, a tiny dose of physics: Absolute zero (−273.15°C or −459.67°F) is the temperature at which molecular motion stops. It is the theoretical lowest limit of temperature. Like the speed of light, absolute zero can be approached closely but cannot actually be reached, as to reach it an infinite amount of energy is required. This requirement is imposed by Heisenberg's uncertainty principle, which is the cornerstone of quantum theory. The German physicist Werner Heisenberg determined in 1927 that it is impossible to find out exactly the position and momentum of a particle simultaneously. This means a particle cannot stop to a standstill at absolute zero; if it does we would know its position and momentum both precisely and simultaneously. At absolute zero it must still vibrate and therefore must possess certain amount of energy. This energy is known as zero-point energy.

So, the concept zero-point energy is absolutely scientific, but not the ability to harness it. Many claims have been made to tap this energy, but no perpetual-motion (*story 67*) or free-energy machine has ever been found to produce a net gain in energy. The laws of physics that allow zero-point energy also limit its behaviour. There is no free lunch, as they say, and there is no free energy.

Some 'energy healers' claim that their zero-point wands or other devices provide a field of zero-point energy to allow the body to remember what it was like at zero point, which causes the cells in the body to reconnect with universal energy. Quantum of science or quantum of nonsense? You decide.

Zoo hypothesis

101
Alien big brother is watching us all

EXTRATERRESTRIAL INTELLIGENT LIFE is widespread. Their reluctance to interact with us can be explained by the hypothesis that they have set aside our planet as part of a wilderness area or zoo.

This is the controversial and demoralising hypothesis proposed by American astronomer John A. Ball in 1973. The hypothesis is based on three premises:

- *Whenever the conditions are such that life can exist and evolve, it will.* Bell believes that the discovery of primitive life on Mars or anywhere else would probably solve this question.
- *There are many places where life can exist.* Since Ball proposed this hypothesis, the discovery of extra-solar planets supports this premise.
- We are unaware of 'them'.

Ball considers the third premise extremely significant. We are not aware of them because they are deliberately avoiding us and they have set aside our planet as a zoo or wilderness sanctuary. In a perfect zoo, animals would not interact with, and would be unaware of, their keepers. 'The hypothesis predicts that we shall never find them because they do not want to be found and they have the technological ability to ensure this,' he says.

He admits that the hypothesis is pessimistic and psychologically unpleasant: 'It would be more pleasant to believe that they want to talk with us, or that they would want to talk with us if they knew we are here.'

In a comment made in 1980 on the zoo hypothesis, Ball suggests that the civilisation that is number one exercises control and enforces the rules: 'They may keep us separate from our neighbours to prevent unfavourable or disastrous interaction.' Or perhaps they are waiting until the time is right to invite us to join the Galactic Club.

German astrophysicist Peter Ulmschneider also supports the zoo hypothesis. He said in 2003, 'While at first sight this idea appears quite extravagant, it makes considerable sense on closer inspection.' He suggests that our history would suffer drastic and fundamental changes – much more so than the history of the native Americans, annihilated when Columbus and Cortez arrived – if it came in contact with an alien advanced civilisation. 'It would be a catastrophe, a culture shock, and essentially an irresponsible act on behalf of the extraterrestrials,' he warns.

Ulmschneider, however, believes that a civilisation as capable of irresponsible acts as human civilisation has been in the past would not survive thousands, millions or billions of years without falling victim to the dangers of such behaviour. His conclusion: if highly advanced extraterrestrial civilisations exist, they have learned to act responsibly. This means 'under no circumstances will they disturb us or contact us, nor will they allow us to trace them by radio waves, through artefacts, or by direct contact'.

This rule out sending flying saucers to check on our wellbeing. Not a happy idea for UFO fans to entertain. Perhaps there is a sign somewhere at the perimeter of the solar system warning all trespassers: 'Wilderness Sanctuary: Don't Enter.' UFO fans can take solace in the knowledge that some cheeky alien teenagers do occasionally ignore this warning and fly their space buggies in our protected skies.

Ulmschneider is not so pessimistic about the future of the human race. He believes that one day we will have the knowledge to discover extraterrestrial intelligent societies in our own galaxy and other galaxies, but the possibility of ever directly interacting with them is very bleak. He is hopeful that, by that time, the basic idea behind the zoo hypothesis – that of acting responsibly – will become the guiding principle for our own behaviour. Moral: When we do escape from our zoo, we should become keepers of other zoos.

Appendix 1
Separating the wheat from the chaff

IT'S EASY TO differentiate between a scientist and a crank: when a scientist proposes a hypothesis he or she tries to find out whether it is true; a crank tries to show that it is true.

A scientist's method of discovery is a continuous interplay of observation and hypothesis: observations lead to new hypotheses, which guide more experiments, which help to change existing theories. With some exceptions, the scientific method involves the following sequence:

- Observations and search for data
- Hypothesis to explain observations
- Experiments to test hypothesis
- Formulation of theory
- Experimental confirmation of theory
- Mathematical or empirical confirmation of theory into scientific law
- Use of scientific law to predict behaviour of nature

Pseudoscience is ideas and beliefs which masquerade as science, but have no or little relationship to scientific method. Theories of real science are continually being added to and updated, but the ideologies of pseudoscience are fixed.

In his 1952 book, *In the Name of Science* (which was republished in 1957 as *Fads and Fallacies in the Name of Science*), Martin Gardner, a well-known author of numerous books and a relentless fighter against pseudoscience who died in 2010, launched the modern sceptical movement. In this book, he lists five characteristics of pseudoscientists:

- They consider themselves geniuses.
- They regard other scientists as ignorant blockheads.
- They believe themselves unjustly persecuted and discriminated against because recognised scientific societies refuse to let them lecture and peer-reviewed journals ignore their research papers or assign them to 'enemies' to review them.
- Instead of sidestepping the mainstream science, they have strong compulsions to focus on the greatest scientists and best-established theories. For example, according to the laws of science a perpetual motion machine cannot be built. A pseudoscientist builds one.
- They often write in complex jargon, in many cases using terms and phrases they themselves have coined. Even on the subject of the shape of Earth, you may find it difficult to win a debate with a pseudoscientist who argues that Earth is flat. Simply put, a pseudoscientist believes that their hypothesis can never be wrong, but a real scientist always welcomes new ideas as these ideas give them the opportunity to test their hypothesis in new situations.

How do you explain a new theory? Some will take you through a maze of facts, hypotheses and observations to arrive at their theories and you will still be doubtful. Others will cut to the chase and direct you convincingly to their theories.

William of Ockham, a Franciscan monk who lived in the 13th century in England, developed a bright rule, now known as Ockham's razor, which is of vital importance in the philosophy of science even today. The rule – it is vain to do with more what can be done with less – implies that the number of causes or explanations needed to account for the behaviour of a phenomenon should be kept to a minimum.

It is a guiding principle in developing scientific ideas, and it insists that you should prefer the simplest explanation to fit the facts. The rule has been interpreted now to mean that when you have two competing theories that make exactly the same predictions, the one that is simpler is better. In other words, the explanation requiring the fewest assumptions is most likely to be correct.

The advice 'keep it simple, stupid' is in a similar vein. But everything should be made as simple as possible, but not simpler. Trim fat, but leave flesh on the bones of your idea.

Ockham's razor is an important tool to have in your toolkit if you would like to think like a sceptic. It can help you choose between possibilities. A sceptical thinking would help you to separate the wheat (science) from the chaff (pseudoscience).

Appendix 2

Why are we suckers for weird beliefs?

ALIEN ABDUCTION, ANCIENT astronauts, astrology, crystal healing, ESP, magnetic therapy, quantum healing ... the list of absurd things is long, and the number of people who not only believe in such rubbish but vigorously defend them is very large. There is absolutely no evidence, scientific or otherwise, to support their beliefs. Ockham's razor – the simplest explanation is most likely to be right – is alien to believers in UFOs; science is crystal clear to practitioners of crystal healing.

Why does the human brain allow and even encourages beliefs that defy reason? The answer probably lies in your parietal lobe – a mass of tissues at the top of the brain which process sensory input and distinguishes where the body ends and the material world begins. During intense prayer or meditation, the parietal lobe powers down. Unable to find the dividing line between self and the world you experience the sense of having lost your worldly moorings. You feel connected to the 'other world'. The parietal lobe's ability to go quiet may encourage other beliefs that bring a sense of connection.

Brain imaging these days can reveal many mental acts. A trio of American neuroscientists, Sam Harris, Sameer Sheth and Mark Cohen, just did that to find out how the brain differentiates between belief and disbelief. Their simple and ingenious experiment involved presenting a series of written statements to participants while they were in the fMRI scanner. The statements were designed to be clearly true (belief), false (disbelief) or doubtful were from seven categories: mathematical, geographical, semantic, factual, autobiographical, ethical and religious; for example:

- 62 can be evenly divided by 9
- Senegal borders Guinea

- 'Devious' means 'friendly'
- Most people have 10 fingers and 10 toes
- You have two sisters
- It is bad to take pleasure at another's suffering
- There is probably no actual Creator God

The results were fascinating. Response time to true statements was much shorter than responses times to both false and doubtful statements; however, there was no difference in response times between false and doubtful statements. Response times to acceptance of belief were similar, whether the participants made judgment in the highly emotional areas of ethics or religion or seemingly neutral area of mathematics.

Most strikingly, different brain regions lit up when responding to belief and disbelief statements. True statements activated areas of prefrontal cortex which are thought to play a role in decision-making, memory and fear, while false statements showed increase activity in an area called the anterior insula which helps to process fear, disgust and reactions to bad smells. This finding suggests that there are two distinct brain systems for belief and disbelief. The researchers suggest that belief or acceptance of a proposition as true has a pleasant and rewarding emotional tone. Disbelief or rejection of a proposition, on the other hand, is often associated with a feeling of discomfort and urge to avoid 'untruth'. 'When someone says something you disbelieve, it has a kind of emotional tone,' says Harris. 'Rejecting someone's statement as illogical or incompatible feels like something.'

Believing or doubting something controls our behaviour and emotions. While doubt tends to inhibit action, beliefs make it easier to arrive at a decision and act on it. Michael Shermer, publisher of *Skeptic* magazine, states that once we form beliefs, we maintain and reinforce them through five cognitive biases that distort our precepts to fit belief concepts:

- *Anchoring bias*: Relying too heavily on one piece of information.
- *Authority bias*: Valuing too much the opinions of an authority.
- *Belief bias*: Accepting an argument on the believability of its conclusion.
- *Confirmation bias*: Seeking supporting evidence, but ignoring discomforting

evidence.
- *In-group bias*: Placing more value on the beliefs of our peers.

Should you worry if someone you know believes in weird things? 'It can cause no harm,' says Stephen Law, author of *Believing Bullshit: How Not to Get Sucked into an Intellectual Black Hole*. 'But the dangers are obvious when people join extreme cults or use alternative medicines to treat serious disease.'

About the author

Surendra Verma is a science writer, journalist and author based in Melbourne, Australia. He has published numerous popular science books internationally which have been translated into 13 languages. His recent books include:

The Mystery of the Tunguska Fireball
Why Aren't They Here: The Question of Life on Other Worlds
The Cause of Mosquitoes' Sorrow: Beginnings, Blunders and Breakthroughs in Science
The Little Book of Scientific Principles, Theories & Things
The Little Book of Maths Theorems, Theories & Things
The Little Book of Unscientific Propositions, Theories & Things
The Little Book of the Mind: How We Think and Why We Think
Learn & Unlearn: The novel way to rethink the things that matter in your life
Science in 100 Words

+ a children's book
Who Killed T. Rex?: Uncover the mystery of the vanished dinosaurs

Visit www.surendraverma.com for more information on these books.

If you enjoyed reading this book, you'll also enjoy ...

SURENDRA VERMA

The Mystery of the Tunguska Fireball

www.ingramcontent.com/pod-product-compliance
Lightning Source LLC
Chambersburg PA
CBHW050207230526
45470CB00001B/278